Dry River

I0887531

Dry River

Stories of Life, Death, and Redemption on the Santa Cruz

Ken Lamberton

The University of Arizona Press
Tucson

The University of Arizona Press
© 2011 Ken Lamberton
All rights reserved

www.uapress.arizona.edu

Library of Congress Cataloging-in-Publication Data
Lamberton, Ken, 1958–
Dry river : stories of life, death, and redemption
on the Santa Cruz / Ken Lamberton.
p. cm.
Includes bibliographical references and index.
ISBN 978-0-8165-2921-6 (pbk. : alk. paper)
1. Santa Cruz River Valley (Ariz. and Mexico)—History, Local.
2. Santa Cruz River Valley (Ariz. and Mexico)—Description and travel.
3. Nature—Effect of human beings on—Santa Cruz River Valley (Ariz. and Mexico) 4. Natural history—Santa Cruz River Valley (Ariz. and Mexico)
5. Human ecology—Santa Cruz River (Ariz. and Mexico) I. Title.
F817.S33L36 2011
979.1—dc22 2010039342

Manufactured in the United States of America on acid-free,
archival-quality paper containing a minimum of 30 percent
post-consumer waste and processed chlorine free.

16 15 14 13 12 11 6 5 4 3 2 1

For Karen

Jessica, Kasondra, and Melissa

—like streams in the desert

Contents

Maps

Illustrations

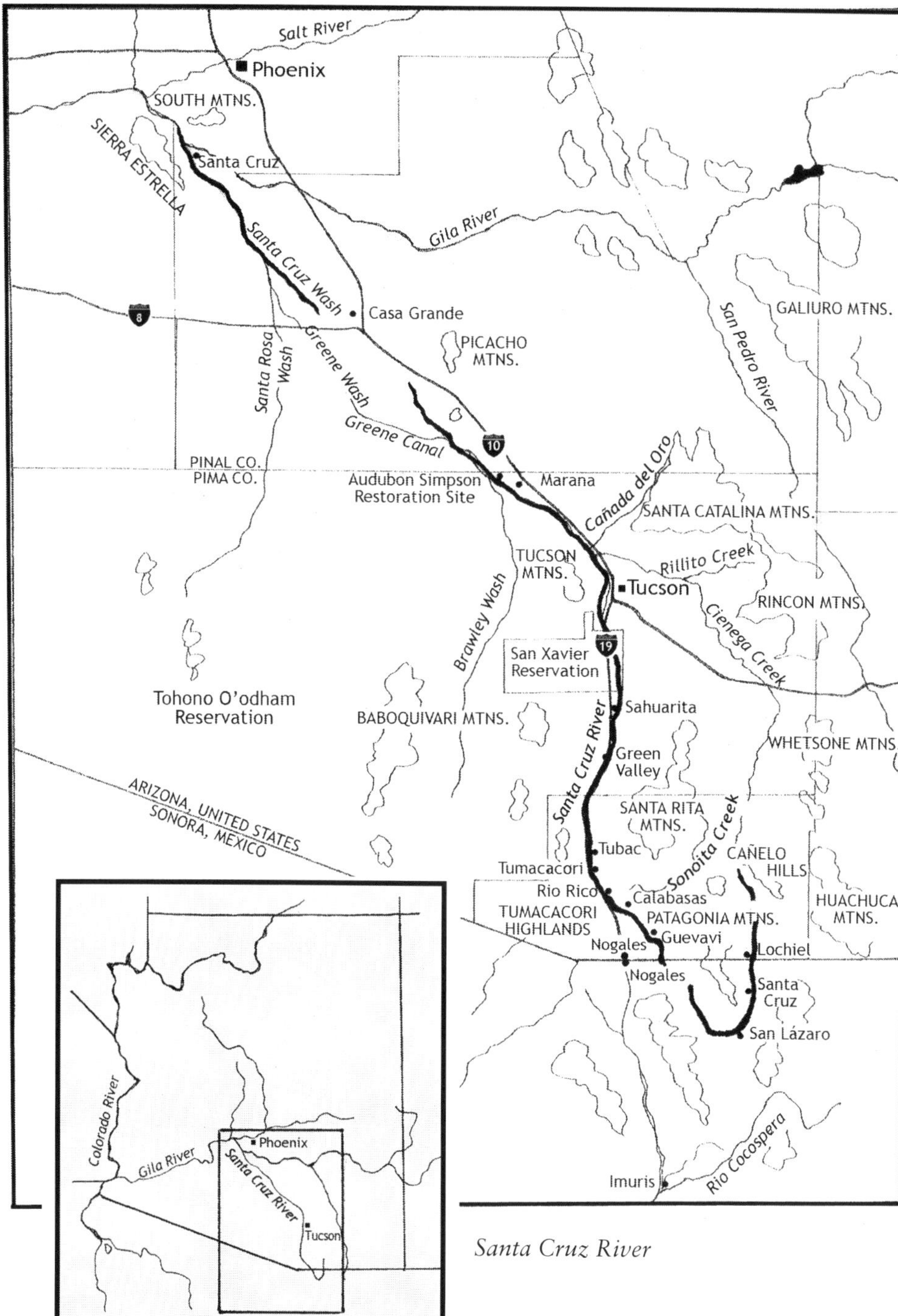

Santa Cruz River

Dry River

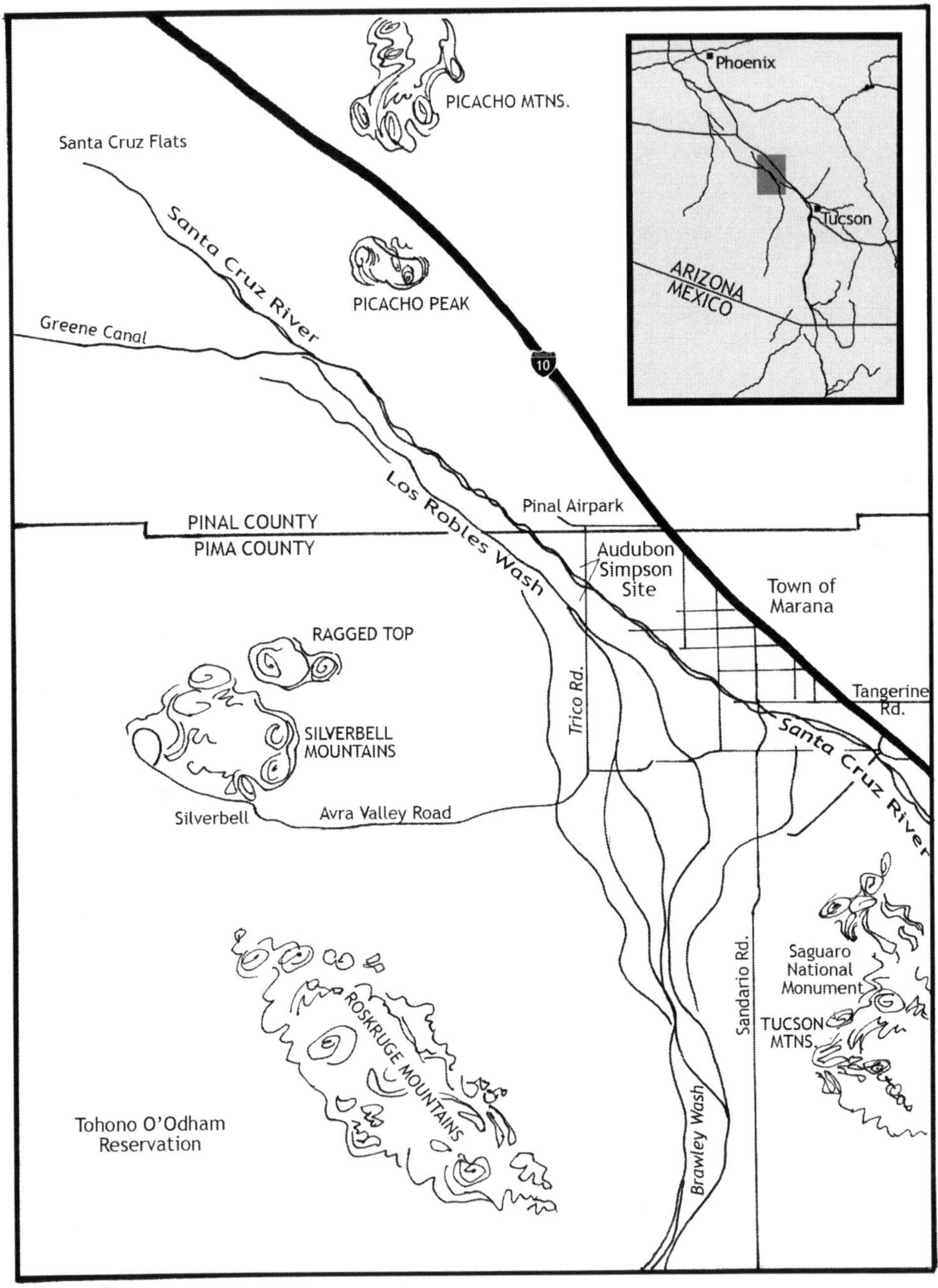

Audubon Simpson restoration site

1

A River Once More

In the desert, one finds the way by
tracing the aftermath of water . . .

—Alison Hawthorne Deming,
Edges of the Civilized World

On the afternoon of the winter solstice in 2002, Daniel Preston and Renee Red Dog of the Tohono O'odham Nation stand in the rain on an earthen berm above the Santa Cruz River thirty miles north of Tucson. The smell of burning sage mingles with the tonic scent of wet creosote. At one of the first water-harvesting basins created at the site, fifteen observers, jacketed and sweatshirted against the December chill, huddle around a potted blue paloverde sapling. As Daniel blesses the tree in the O'odham language, a worker slips it into the ground.

Next, the group climbs down the soft bank to the river. Daniel, wearing his trademark bolo tie, speaks about the need for a blessing in this place, how his ancestors once drew life from the Santa Cruz and how life has begun to return here now, coming full circle. He asks everyone to face east, and then he and Renee begin to sing, praying for strength and guidance for the people who are working diligently to heal the river.

For the thirty-five years since I came to this desert, the Santa Cruz River has been a mystery to me. It is unlike any river I have ever known. But, I wonder, does the river really need healing? Haven't the dramatic end points of this strange and twisting gash always been this way? River of still and silent sand in one moment; river of thundering mud the next. The desert, I've come to learn, is an extreme place. So maybe this dry river that runs through it has to be the desert's reflection.

I envision my feet sinking deeply into river-bottom sediments, black ooze squeezing up between my toes, smaragdine algae streaking my shins. The water is cold. Its current pulls at each insistent step as if the water itself were imbued with the qualities of a living thing. This is the kind of river I know. The other southwestern channels, more liquid earth than water. The Rio Grande, where you slip to your thighs in mud and barely escape the embrace. The Gila, with its peristaltic current that drags you through overhanging mesquite trees and into its smothering depths. The Salt, with its ferocious gorges and white water. The Colorado, whose suspended grit of seven states fills your pores and polishes the enamel from your teeth. Even the San Pedro still flows beneath a forest of trees.

But the Santa Cruz—if it's imbued with anything living, it keeps it to itself. The river is slow to reveal its nature, that part of it that inhabits the space between the extremes. Between its poverty and richness, the river is hardly an afterthought. What does water mean to the Santa Cruz? It certainly doesn't define it, since over most of its course water is an abstraction, or at best only wishful thinking. The Santa Cruz River *is* the aftermath of water.

This is the mystery. Yet I have to believe that there is more here, more than aftermath. And that I can find it and maybe just begin to understand it, even if its measurement lies beyond what human timepieces can record and must be measured by the chronometer of deep time. This is the way of *this* desert river, as it always has been, and will continue to be: Droughts, floods, rising and sinking aquifers—and they all lie on a continuum that is the Santa Cruz River. My presence here hardly fills the space between cosmic ticks. But it is enough for a beginning.

Ten months after the blessing ceremony, my wife, Karen, and I return to the lower Santa Cruz River to search for the "hobbit hole," a name our fifteen-year-old daughter gave to her planting design. Melissa had an idea that her plants should benefit from both rainwater and the provided irrigation and that the inevitable weeds shouldn't interfere with either. The result was a cloverleaf of six catchment basins, each one banked and sloped to receive and hold the greatest amount of moisture while channeling it to the roots of her seedlings—a yucca, a creosote bush, a saltbush, a blue paloverde tree, and two bunchgrasses.

We worked together for most of the day last December, our faces burnishing red from sun and exertion. It wasn't the most efficient use of our time—six tiny plantings in the center of an abandoned cotton field where volunteers were digging hundreds of holes in sandy loam the

The author and Melissa Lamberton at the Audubon Simpson restoration site hobbit hole

color and texture of cocoa powder. But it was our project, our small part toward restoring seventeen hundred acres of tumbleweeds to a habitat more suitable for wildlife along the lower reaches of the Santa Cruz River.

Now, the weeds have grown tall, Karen and I notice, and they obscure Melissa's hobbit hole. At first I think that nothing has survived, but then I see a small yucca, blue-green against the dead-blond vegetation. Alive. The creosote is more than twice its original size, and even the paloverde has new growth.

Ann Phillips, the project manager for the Audubon Simpson restoration site, tells us that she's refitted the original irrigation system with drip lines rather than sprinklers. "We learned a lot at this site," she says.

Ann gave us a tour of another area of the site earlier this morning, after Karen and I helped a group of new volunteers, the first of the season, plant dead trees on a berm above the opposite side of the river. "We're having trouble with rodents chewing on the irrigation lines," she explained. "The dead trees will give hawks a place to perch and deal with the rodents."

Ann has been involved with the Santa Cruz River Habitat Project since 1999 when she worked as a contractor for the Tucson Audubon

Society. Her ideas are innovative. She's a visionary. She sees possibility in barren fields.

On some used-up fields along this part of the Santa Cruz, Ann brought a team of conservation biologists to create a site assessment that mapped the local characteristics like topography, soil type, and temperature. With this information, together with historical records and input from botanists, she selected plants appropriate for the place. The result was a list of sixty-nine species of trees, shrubs, cacti and succulents, wildflowers and grasses, which she divided into plant guilds according to species that grow well together. For example, Melissa's paloverde tree would both fix nitrogen in the soil and provide cover for the saltbush, creosote, yucca, and grasses, which in turn would offer shelter and food for birds and mammals.

Besides constructing water-harvesting basins, another of Ann's innovations is what she calls "assisted suicide for tumbleweeds." She stumbled on the technique accidentally after noticing how piles of dead tumbleweeds suppressed the growth of new tumbleweeds. Since the plants spread their seeds once they dry and begin dashing across the desert, removing dead plants by hand only served to scatter more seeds. Ann discovered that crushing the plants in place (by laying wire fence sections over the plants and jumping up and down on them) would choke out the following year's tumbleweed sprouts and at the same time provide mulch for the already thriving native species.

I learned about the Santa Cruz River restoration project from the *Vermilion Flycatcher*, the Tucson Audubon Society's monthly newsletter, which ran an article about the program and the need for volunteers. For several years, I had been considering writing a book about the Santa Cruz River, hiking through the river's natural and cultural history from its source to its cessation, focusing on current restoration work taking place along its course. I would weave in personal stories about my family, how the Santa Cruz has affected our lives and other people's as well, past and present. The book would be about redemption and hope, rather than abandonment and despair, themes already more than adequately addressed concerning the river.

As mentioned earlier, my connection to the Santa Cruz River began at childhood upon my arrival to southern Arizona. Like the wildlife, I relied on the river channel and its leaf-veined tributaries as a corridor for migrating through a heat-blanched landscape, although this was more about exploration and discovery than about finding food and suitable habitat. For me, washes with names like Santa Cruz, Rillito, and Cañada del Oro were magical realms of brush forts and tadpole-black pools, places to hunt for speed-blurred whiptail lizards and gopher snakes to keep in my pockets, places to tunnel into coyote dens and breathe in the root-smell of dirt hidden from sunlight for centuries, places to became intimate with the desert.

Even today, as I write these words, a winter storm washes away sediments I walked barefoot over this morning, the flooding arroyo behind my home carrying my skin cells and footprints to the Santa Cruz, *connecting* me to the river in a profoundly personal way. It seems I couldn't avoid writing a book about the river. My words are a natural expression of this connection, although I had little idea where the connection would take me, in my writing, in my life. Rivers are like that.

So, because of my interest in writing about the river, and particularly because I have a heart for redeeming hopeless cases (I regularly burn a candle to St. Jude in my window), I decided to participate in the restoration project and to bring along my family, if Karen and our three daughters, Jessica, Kasondra, and Melissa, were willing.

On a Saturday morning in October, my oldest daughter, Jessica, and I joined Ann Phillips, two of her staff, Kendall Kroesen and Rodd Lancaster, and a group of volunteers at some abandoned cotton fields near Pinal Airpark. The farmland, now owned by the city of Tucson, had been retired since the late 1970s.

In recent historical times, surface flow from the Santa Cruz never provided much in the way of irrigated agriculture below the Tucson Basin, where the river remained dry except during heavy storms and floods. Occasionally, workers at the restoration site unearth a few shards of Hohokam pottery, indicating some prehistoric activity along this part of the river, but beginning about three hundred years ago the region's climate had dried significantly enough to wither any crops dependent on irrigation. The Tucson Farms Company made an attempt at farming the lower Santa Cruz in the early 1900s. Investors poured millions of dollars into the Santa Cruz Reservoir project, a series of dams, catch basins, canals, and levees designed to channel and hold what little seasonal runoff the river and its tributaries could muster. Greene's Canal, an ambitious thirteen-mile-long component of the project named for Colonel William C. Greene, a colorful Arizona-Sonora farmer turned miner turned rancher and gunslinger and the canal's principal architect, connected the Santa Cruz River to a reservoir from which farmers could draw water for their fields. The project failed in the end—due mostly to high evaporation rates in the reservoir—but not before floods in 1914 and 1915 destroyed a diversion dam and altered the course of the river. Today, the lower Santa Cruz River no longer follows its former, northerly, path where the town of Casa Grande now lies. Instead, because of the wide arroyo of Greene's Canal (Greene's Wash), it joins a collection of drainages on a more westerly route to the Gila River.

Successful agriculture finally came to the area with the advent of groundwater pumping. Where the river proved inadequate previously, now a seemingly limitless aquifer could be tapped. Ancient and hidden waters once squandered by porous sand could be returned to the roots of a new civilization, all with the proper application of industry and ingenuity. By the mid-1930s, groundwater supplied the irrigation for crops along the Santa Cruz River from the Rillito confluence to the Gila River, over ninety-eight thousand acres of new farmland. Then, with the price of cotton on the rise due to the Second World War, farmers redoubled their draw on the aquifer. By 1949, pumping exceeded a total of one million acre-feet of groundwater. In 1952, it was ten million. Like earthworms turning a garden inside out, more than a thousand wells lifted the aquifer and spilled it onto the desert, and the desert began to sink. In some places during the next three decades the water table fell by fifty feet, the land surface by fifteen feet. Although much of the subsidence was gradual, near Picacho Peak a fissure known as "El Grande" unzipped a seam hundreds of feet wide and ten miles long.

Tucson's interest in the lower Santa Cruz began in the 1960s as its population approached three hundred thousand. The city no longer cared about agriculture; in fact, agriculture within the Tucson Basin had already become a threat because of the competition for water. Cotton could be imported from elsewhere. Now, the city needed to tap the fringes, specifically the aquifers beneath the upper and lower Santa Cruz River. The result, amid a barrage of sometimes messy water politics, was that Tucson bought the farm, taking cotton fields like these at the seventeen-hundred-acre restoration site out of production for the water that pools beneath them.

The warm air smelled of dust and effluent, the latter unfurling in a dark liquid ribbon along a cottonwood- and willow-hemmed seam

Audubon Simpson restoration site

immediately to our northeast. The channel is human-made, we learned, bulldozed after the flood of 1993 stripped the region clean of vegetation and left behind a plain of sand and mud. Since engineers reclaimed the river channel, cottonwood and willow, mesquite and paloverde, with invasive trees like tamarisk, have sprouted and surged into thirty-foot galleries at the river's margins while the silted fields became a cropland of tumbleweeds. Here, because of two sewage treatment facilities upriver, the Santa Cruz flowed once again, the river given a second chance by Tucson's unwanted wastewater. As a result, the whole region was ripe for restoration.

That Saturday, I worked with Jessica along the riverside berm, digging holes to sow a little redemption in my own landscape. We enhanced natural water basins for acacia, brittlebush, four-wing saltbush, and sacaton grass grown in two-foot sections of four-inch PVC pipe by Mountain View High School students. Kendall, in beard and khaki Nature Conservancy hat, explained that the tubular "pots" encouraged long taproots, which improved the seedlings' survival rate. "The site gets only about ten and a half inches of rain a year, about a third of it coming in heavy summer thunderstorms. Wide basins and long roots help the plants take advantage of the moisture."

Kendall Kroesen is the right arm of the restoration project. Along with his duties as editor of Tucson Audubon's *Vermilion Flycatcher*, he is a permaculture specialist whose handiwork at the site appears in mesquite-clotted basins and swales. His background is in anthropology, having received his PhD from the University of California, San Diego, in 1997. But his interests in people and nature, and his research into what makes communities successful, have led him to seek ways to create a more sustainable human society, particularly in the Sonoran Desert where he now lives. Kendall sees ominous trends in today's commerce-driven society, which consumes nonrenewable resources along with cultural diversity and widens gaps between the haves and have-nots. "I would like to explore social forms that allow for maintenance of natural ecosystems and threatened wildlife alongside humanity," Kendall says, which is why he's so deeply passionate about this restoration work.

On a later trip to the river restoration site, my whole family joined me. In March, with temperatures near seventy-five degrees, Jessica, Kasondra, Melissa, and I worked the berm, flattening tumbleweeds, digging up Johnson grass, and planting mesquite trees. Karen took photographs of the staff and volunteers. With the sun warming my winter skin and the taste of the land on my lips, it felt good to raise blisters with family and friends.

Now, in this new season, Karen and I stand next to the hobbit hole admiring one small effort to heal the land. Daniel Preston's ancestors knew how to cooperate with the Santa Cruz to make this extreme place into a human landscape, something modern civilization could never do with all of its giant canals and aquifer pumps. This region is a lesson in extremes; travelers here a hundred and fifty years ago called this a waterless "Ninety Mile Desert" and write of trails marked by wagon graves and cairns of human bones. The stories remind me that we live in the desert by grace, and that our presence becomes more tenuous as the rivers go.

It is significant to me that our restoration work of the Santa Cruz should begin here, some distance before the river finally expires, dilating and sinking into the earth. The river, which has never been much more than a meager, disjointed stream—perennial or ephemeral in some places, dry in others—has had a tough life. From its headwaters in southern Arizona's San Rafael Valley, its journey through Mexico, its return to Arizona and subsequent use and abuse by the state's formerly two largest economies, to the cutting of its canopies and erosion of its banks and its treatment as a landfill and sewer, the Santa Cruz has suffered. And yet . . .

A gray-feathered hawk turns ragged circles above our heads, over a sere land slashed open by a bright green invitation. Desert rivers are like this, gushing with extravagant promises that evaporate before your eyes. Oases without water on their way to becoming oases without trees. Once, the Santa Cruz was one of southern Arizona's major arteries, a place to take the desert's pulse. It could be again. All we need is hope—hope because, unlike despair, hope calls for action. It means another world might yet be possible.

I think about redemption, what it means in terms of people's patience and sacrifice and dedication, and I consider the possibility that we can find the way by following the aftermath of water. I imagine how this place might look in fifty years. I see a mesquite bosque crowding the margins of a living thread—a free-flowing Santa Cruz River—connecting wildlife and habitats for hundreds of miles of Sonoran Desert from Mexico to central Arizona. I see into the past, the way the river once was. And at the same time I see into our future.

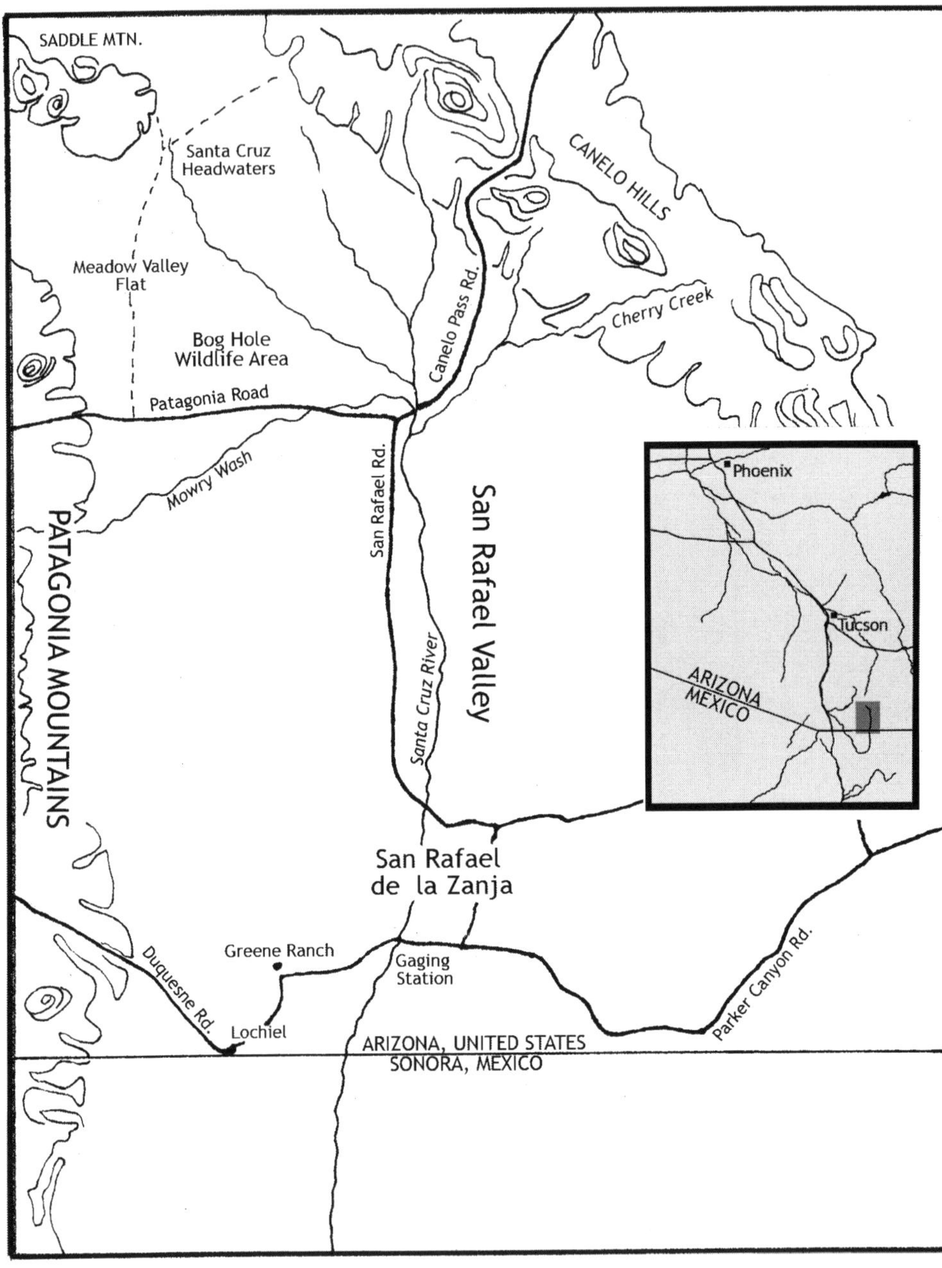

The San Rafael Reach

2

Sources

The San Rafael Reach

Headwaters: February, 0 River Miles

Karen and I drive over Canelo Pass and suddenly San Rafael Valley drops away to the southwest. Before us opens a broad rolling plain, crisscrossed with ropes of green where oak and juniper line wrinkled drainages. The wind cuts northeast, strumming a rhythm in knee-deep grasses as blond as Karen's hair; the bent stems sound like rushing water. Mountains shoulder the valley—the Huachucas on the east, Patagonias on the west—their peaks blue-gray above tan-flecked flanks. Clouds of multiple configurations slip out of Mexico to lay stains across a rangeland where white-crowned sparrows call from beyond barbed-wire fences.

On my topo map, the headwaters we're looking for lie across the upper San Rafael Valley like a giant cottonwood leaf, its veins all tributaries flowing into the leaf's petiole where the gathering river spills south into Mexico before changing course and returning to Arizona. We're seventy-four miles from our home in Tucson, the place we last crossed this dry river on its way north. I park our minivan in a rock-studded drainage. A weathered sign reads: "Santa Cruz River."

We both look "upstream," and Karen says, "Let's follow it from here." But I know from the map that it must still be half a dozen miles to the headwaters, so I drive on.

Near Saddle Mountain, northwest of the valley, the two-track I'm negotiating slips under a fence and vanishes in the grass. We get out to walk. Somewhere I've read that the "official" source of the Santa Cruz River lies beneath this fifty-eight-hundred-foot granite outcrop, but I'm not sure where, and I haven't bothered to ask anyone. The river should be on my right, but the whole landscape drops sharply away on my left, forming a canyon country whose ridges and drainages corrugate the southern flank of Saddle Mountain and sweep southward into the Patagonia Mountains.

Karen and I stay on the high plain, swishing through the grass or taking one of the beaten paths used by Mexican immigrants identifiable by the sunken bags of Bimbo bread and torn cans of tuna we find along the way.

Mexicans are only the most recent wave of immigrants here. Humans probably first trickled into the San Rafael Valley at the end of the Pleistocene between ten thousand and twelve thousand years ago when groups of hunters now known as Clovis people stalked now-extinct megafauna across this landscape. Only thirty miles to the east of here, at a place called Murray Springs in the valley of the San Pedro River, erosion has exposed a butchering camp where Clovis hunters eleven thousand years ago killed an adult female mammoth and a dozen bison.

Looking southward into the past from the head of this valley, its entire length swept with dappled cloud-shadow, I imagine nomadic bands of Paleo-Indians following the folded-grass trails of fattened mammals. As the Ice Age drew to a close and the land relaxed under warmer skies, these interconnected valleys like the San Rafael with their high stalks of wild grains may have helped these people settle into a more hunter-gatherer lifestyle, and possibly even form communities that shared social and religious customs.

At the Lone Mountain Ranch, ten miles east of the river in the southern foothills of the Huachuca Mountains, anthropologist Charles Riggs and his team have uncovered human occupations spanning more than eleven thousand years. "Through time," he writes in his report about the digs, "this upland zone appears to have formed a special ecological niche, first for mobile groups and later for sedentary groups using or occupying the lower riverine and grassland environment." Today, a gerrymander of oak woodlands and grasslands drapes hills cut by riparian drainages with names like Joaquin Creek, Bear Creek, and Cave Creek, ephemeral streams reduced in the dry season to chains of stagnant pools. Lone Mountain Ranch is a museum display-case of prehistoric artifacts: Boulders show etchings of human figures, lizards, double helixes, and spirals; fire-cracked rocks circle roasting pits; smooth fists of stone and grinding metates litter the ground among flakes of chert, mudstone, and obsidian, decorated and unpainted pottery sherds, and small, triangular Elko and Pinto projectile points.

The San Rafael Valley holds a long, nearly continuous presence of humans. Ten thousand years ago, people of the Cochise culture shared the grasslands with fantastic animals like lions and camels, and 110-pound long-fanged dire wolves, camping here to harvest wild grains and parch them in hearths of incandescent coals. Three thousand years ago, the area had been a home to people who painted red crosshatches and parallel lightning bolts onto their smoky brown pottery and roasted agaves in the ground to feast on their dark and sweet molasses-like cores. These people probably traded with Hohokam in the north, Mogollon in the east, and Trincheras to the south and may have been Hohokam themselves, but this is still open to question according to Riggs. "They are in a sort of transitional area between Hohokam and Mogollon (San Simon Branch)," he told me recently. "But there was a fair amount of Santa Cruz red-on-brown [pottery sherds], which would suggest Tucson Basin Hohokam." Later, during the first millennium AD, a hamlet of brush-and-mud pit houses provided homes for a community of farmers.

For thousands of years, this place has drawn people, the latest being four nineteenth-century homesteads and a cattle ranch. What all these people have in common is their connection to the valley, to the resources it provides, and to the river that sustains them. In fact, the entire Santa Cruz River, from the headwaters in this valley to the Tucson Basin and beyond, braids together the footpaths of humans dating from prehistory forward. It's as if the river made it possible for people to find their way in this world, and to find a way of being in this world.

And it still does today.

On the windowsill above my desk, a rough pentagon of broken pottery rests beside other accumulations—a devil's claw, coral bean seeds, pieces of rhyolite, a dried papery slip of desert spoon—from the desert outside our home. Karen's mother found the sherd recently in the backyard arroyo. It is thin and slightly curved, like a piece of eggshell, and across its convex surface runs a tracery of dark parallel lines. Black-on-white. It looks like Hohokam.

This arroyo bisects five acres of desert that Karen has lived on since her birth. Called the San Juan Wash, it runs east into the mesquite-clotted West Branch of the Santa Cruz River just over a mile away. Her whole life has been connected to the river. (Her father still talks about fishing in the Santa Cruz from the Ajo Way bridge when he was a teenager in the 1940s, a bridge we take on our daily trips to and from the city, a bridge that now spans only sand. So she's a second-generation Santa Cruz aficionado.) For more than forty years, Karen has lived and played in this drainage, hiking its dry course through burgeoning hills stabbed with saguaros or soaking her feet in chocolate runoff after a monsoon thunderstorm. Her cells, along with the soil and rock and sherds of Hohokam pottery, have tumbled down the San Juan Wash and poured into the Santa Cruz River, churning and foaming together with the essence of this place and its people as water becomes life. Karen *is* this river.

Today, we live together on the land of her beginning, having raised our three daughters on the same desert gouged by an arroyo named for St. John the Baptist that empties into the river of the Holy Cross.

Against her better judgment, Karen follows me into the western canyons beneath Saddle Mountain. I'm thinking that this wild canyoned landscape must drain the headwaters, collecting the young Santa Cruz to our south before directing the waters eastward into one of the blue-lined squiggles on my map.

Saddle Mountain is a 65-million-year-old granite intrusion that formed as magma cooled slowly beneath the surface and later rose skyward as the adjacent terrain slumped into canyons and valleys. As wrinkles betray a person's age, canyons reveal the history of the earth, in this case seams of sedimentary deposits and lava flows laid down eons before more recent tectonic events wracked the region and exposed the buried rocks. About 17 million years ago, these new forces began stretching the western United States, breaking the crust into huge sections that shifted like an unbalanced teeter-totter, tilting up whole mountain ranges and dropping wide valleys in parallel trackways. During the interim quiescent periods, the basins filled with mountain debris, creating a far-reaching landscape we see today: the Basin and Range Province.

Author John McPhee says this recent mountain building has unearthed the "chaotic, concatenated shards of time." "In the Basin and Range," he writes in his book of the same title, "are the well-washed limestones of clear and sparkling shallow Devonian seas. There are dark, hard, cherty siltstones from some deep ocean trench full of rapidly accumulating Pennsylvanian guck. There are Triassic sediments rich in fossils, scattered pods of Cretaceous granite, Oligocene welded tuffs. There is not much layer-cake geology. The layers have too often been tortured by successive convulsive events."

Unlike most mountains ranges, which form due to pressures that compress and bend and fold the earth's surface, the mountains of the Basin and Range Province result from an expansion of the crust, a northeast to southwest spreading that rips and tears on a northwest to southeast line. Here, the world is coming apart at the seams. Hundreds of long, narrow, "fault block" mountains separated by wide basins slash obliquely across the West between the Sierras and Rockies from southern Oregon and Idaho through Nevada and Utah into southern Arizona and northern Mexico. These are the stretch marks of our present layer of landscape. It is a pulled-apart and fractured, lifted and sunken region that lays bare the landscapes of eons, as the earth's crust explores fantastic rhythms to accompany the mantle's fluid harmonics in what is truly music of the spheres.

During the early Pleistocene between one and two million years ago, shallow runoff poured into the intermountain basins, charging aquifers and depositing gravel and sand alluvium from the eroded aprons of the mountains. This was the time of the great inland seas: Lake Manlius of today's Death Valley, Lake Lahontan of the Humboldt and Carson sinks, and Lake Bonneville, progenitor of Great Salt Lake. Farther south, our own Lake Cochise (now Willcox Playa) periodically inundated the Sulphur Springs Valley. These pluvial lakes repeatedly filled and evaporated according to climatic pulses, like those that spurred the wax and wane of Pleistocene glaciers. Then, about five hundred thousand years ago, wetter pulses resulted in the movement of sediments between previously isolated basins and, beginning here in this valley, the formation of a "proto" Santa Cruz River, which probably flowed only south. Over time, the river continued to join basin after basin, raising terraces in one place only to later redistribute them elsewhere. But eventually the river took on its present shape, surging across this land, connecting basins like threading beads on a string until finally the beads slipped the necklace to join a young Gila River on its way to the ocean. Ebb and flow, the recurrent pattern of life, from the blood in our veins, to ocean tides, to planetary glaciations, gave birth to the Santa Cruz River.

We rest in a dry cut under the southeastern flank of Saddle Mountain, where I'm still thinking we might locate the headwaters of the Santa Cruz River. Finding cover from the windchill beneath two large netleaf oaks, Karen pools herself between boulders while I scribble notes about how lime-skin lichens mottle Saddle Mountain's exposed face, about how the grassland slips into several near-gorges like the one we're hunkered down in. Karen reads Mercedes Lackey's *Magic's Price*. Fantasy books are her escape, she explains. Even when she escapes with me.

The first time I noticed her, Karen had just emerged from high school, where books sheltered her from engaging in much of a social life. But sometime near the end of her senior year she must have come to the end of a series, raised her head, noticed her peers, and finally decided to join them. She was seventeen. By graduation, she had purchased fashions from Cele Peterson, styled her hair like Farrah Fawcett, and lost the braces. In a matter of a few bookless months, Karen went from cipher to sexy. And then she met me.

Farther downcanyon, ankle-twisting fists of rock stud the slopes among flourishes of sotol and yucca and ocotillo—so typical of Arizona vegetation at this elevation. Dark outcrops, some smudged with a spectrum of lichens, rim the channel slot. Below us, in the direction we're headed, oaks choke the canyon floor. I'm beginning to suspect that this is all wrong.

When we're a mile into the drainage, late afternoon shadows cling to the steep walls rising on either side of us. We follow a narrow passage to a small seep that pools above a thirty-foot pour-off. Our first water. Karen looks at the algae-rimed scarf of liquid and says: "This place holds only the memory of water."

The sand amasses the tracks of deer, like dozens of empty parentheses, scattered amid the signs of other travelers, such as the boot prints of Mexicans going north. More than four and a half centuries ago, not far from here, possibly even along this upper stretch of the Santa Cruz River, Francisco Vásquez de Coronado and his entourage of soldiers, herdsmen, and herds, made their way north toward the fabled treasures of the Seven Cities of Gold. And still people come, seeking a better life if not the promise of riches.

The migration will never end. Nature teaches that when you have inequality across a permeable membrane, osmosis occurs until there is balance. A fact of nature *and* politics. Since we cannot build an impermeable barrier between us and them (although we might try), we are left with our current feeble strategy of reverse osmosis pumping, which takes significant energy and resources. The answer, I believe, may lie with allowing equilibrium, by opening the gates.

But I also understand the other side, the resistance to helping or encouraging these people even if it means simply leaving bottles of water for them in the desert. Edward Abbey, whose writings I greatly admire, advocated closing the border entirely and protecting it with thousands of heavily armed guards. Abbey, who died in 1989 and was buried secretly in the borderlands he cherished, possibly envisioned the Department of Homeland Security. Without limits to immigration, he claimed, "the social, political, economic life of the United States will be reduced to the level of life in Juarez, Guadalajara, Mexico City, San Salvador, Haiti, India. To a common peneplain of overcrowding, squalor, misery, oppression, torture, and hate." But I wonder what Abbey would say today, after seeing what our immigration policy is doing to his beloved Cabeza Prieta National Wildlife Refuge. I doubt he imagined Hummer tracks over his grave.

At the bottom of the canyon, there's no doubt left in my mind. The drainage flows to the west. This can't be the headwaters of the Santa Cruz.

I search the eastern rim of the canyon and make an estimate as to the position of the car, then trace with my eyes the most direct route out. Karen's not happy about it—both the fruitless chase for the river's source and the climb. She's tired. We both are. "It could be worse," I joke. "It could be raining."

After a half-hour climb, we arrive back at the car. Karen is ready to head for home, but I haven't yet found the headwaters, although I now have a better idea of where they may lie. "Let's just drive around a bit—no more hiking," I promise. "Then we'll go."

I drive back to a fork in the road we passed earlier and turn left, following a northeast track into Meadow Valley Flat, a broad expanse of grassland that slopes away to the east of Saddle Mountain. The road is rough, too rough for our minivan, and I must dodge the regular ruts of hardened mud and eruptions of rock. Where the road pitches downward into a wide seam creased with naked cottonwoods, I give up the effort, back up the road, and pull off into the tall grass to park. "I'll just be a few minutes," I tell Karen, who prefers to stay with the car this time and read her book.

I continue down the road on foot, keeping an eye on the cottonwoods to my right. This must be it, I think. Cottonwoods mean water, even if it's only beneath the surface. They mean life. For generations of travelers in the desert, the trees have represented hope from a distance. Their bright green leaves, flashing their white undersides in the wind, were a beacon in an otherwise inhospitable country.

These cottonwoods are the first, the highest in the drainage, which spills from paired creases below juniper-speckled Saddle Mountain, gathers itself before crossing the road and tearing out a gully beneath the stand of trees. The gully is dry, its grassy margins pinned open with boulders. But I'm convinced this is where the Santa Cruz River begins.

A few hundred yards before the Meadow Valley Flat road joins the road to Patagonia and our route home, the oil light on the dashboard blinks on and remains red. I feel blood drain from my face; I've never before seen this warning light. We're a few miles from the middle of nowhere. We've seen no one, no other vehicle all day. At the crossroads, the engine begins to rattle and then quits altogether. I get out to check, look under the car, and see the oil. It's everywhere. Splashed across the oil pan, streaming from the oil filter, pooling onto the ground. I reach for the filter and feel its crushed form, finger its unseated seal, warped by the perfect impact of smooth rock. I stand, showing Karen my slick and black hands, which say it all, then watch her eyes slide from my hands to the white, mud-garnished 1976 Chevy Suburban driving up behind me.

"Need a hand?" asks a man in a white cowboy hat with the customary western drawl. His name is Zay Hartigan, and he has tools, duct tape, and an amazing six quarts of oil. While we work on repairing the busted oil filter, attempting to straighten the seat and seal it with tape, Karen writes notes in my journal, focusing on a description of Zay. She calls him our rescuer, "a rancher in a truly original work truck," wearing "tall army boots laced up his calves mid way (what I wish I was wearing), a hat, tan leather jacket, and jeans." The jeans, I notice, he's tucked into his boots, which seems practical if not altogether stylish. Karen doesn't mention his tattoos, one of a rattlesnake coiling around his wrist that bared its fangs as he reached to shake my hand.

With oil in the car, Zay tells me that I can coast most of the way to Patagonia—five miles of downhill—except for one small hill, where I should start the engine. Then he says he's off for home where his wife and new baby are expecting him. His plan works until the hill, when the duct tape fails and the car ejects four quarts of oil onto the road in one quick stream. Before we slow to a stop, I see Zay in my side-view mirror pulling up next to us. He has a chain, which he uses to tow us to the crest of the hill and get us coasting again. Once in Patagonia, Zay produces a cell phone and calls a friend of his, Jim Pendleton, who owns Arizona Fabrication and Towing and who might have a new oil filter for us. It's Saturday evening and the place is closed until Monday, but Jim offers to open his shop and find what we need. Zay provides the shuttle service to and from the shop.

I'm impressed with Zay's kindness and I tell him so, but he says it's nothing, that he rescues people all the time on this road, people whose vehicles break down or who lock themselves out of their cars. He's a local celebrity, I decide, after some "gals," as he calls them, drive up in a Model T Ford and recognize him. "Hey, Zay," one of the women asks, "what are ya doin'?" "Just tryin' to help out some folks who got stuck," he says.

With the minivan running again, we exchange addresses. He won't take any money from me, telling me instead that I should pass along the favor to someone else in need.

When Karen and I arrive home later that evening, there's an e-mail waiting for me. It's from Zay. He says, "I hope you got home with no problems, please let me know when you get this mail. I'm off to bed soon, but I'll rest easier when I know."

Over the following months, Zay and I develop a casual correspondence via e-mail. I tell him about our daughters' interests in horses and ranching and he assures me that they would love his place in the valley, and that he and his wife Darlene would love them "if they're anything like you two. Your sense of adventure and good spirit under bad situations is admirable."

Zay eventually shared with me that he was born in Detroit in the late sixties, came west when his family moved east, and followed his thumb all over the country with a skateboard strapped to his backpack. He sent me a photo of himself from those days—naked except for an electric guitar, wearing a Mohawk and a dragon tattoo that lies across his chest and wraps its tail around his lower back and buttocks and slides down his right thigh. He calls the tat Zaharonasaurus; it's one of eighteen that decorate his skin, including his first, which he admits was self-inflicted from a gunshot.

Eventually, he came to Arizona to distance himself from the hard drinking, punk-rock lifestyle he entertained in California. He gave up the alcohol and entered a transformation during four years of what he calls his "Redington phase," named for a small town east of Tucson's Catalina Mountains, where he divided his time between a farm on the San Pedro River and a forty-five-section ranch on the eastern slopes of the mountains. "I was very isolated at that time," he says, "and spent long days on the land, with only horses and dogs for company. That gave me a lot of time to think, as well as a wonderful relationship with living on a natural patch of land."

Zay had been introduced to wildness growing up on his grandmother's farm, and had always planned on coming back to the land. He married Darlene after moving to the San Rafael Valley and describes her as an urban (Chicago) anti-gun type turned "frontier woman," who packs a pistol and drove herself to the hospital to have her baby. "Lucky me," he says, "to have ended up here in this valley for the last six years, and who knows how much longer."

Headwaters Again: October, 0 River Miles

More than seven months later, in early October, Karen and I return to the San Rafael Valley—this time without the minivan, which we've recently traded in for a small, economical SUV (a sign that our daughters are growing up and moving out). I drive along the same busted-oil-filter road, mindful of my undercarriage, and wondering if I'll see a black stain in the dirt. I know Karen is fantasizing about another rescue from her cowboy. She's ever hopeful.

I park where the gully cuts across the road, thinking about what Zay said about the Santa Cruz River having another source, one other than under Saddle Mountain, one where water actually flows all year. This source is dry. Along the gully immediately downstream from the road, reddish-gray boulders lift their swollen haunches above bleached grasses like grazing mastodons. The rocks are an outcrop of worn andesite, testimony to the lava sheet that blanketed the region during the Laramide orogeny, which ended 50 million years ago. Here, at the head of the San Rafael Valley, these ancient feldspar-rich rocks rise near the surface, and the overlying sandy mountain alluvium is shallower, holding water in some places above the ground in springs and bogs.

Karen and I change into hiking boots and fill water bottles, Saddle Mountain's bright thumb at our backs. The grasses are tall, sun-bleached, and their empty seedheads flag with each gust of wind. We follow the gully downstream toward a line of green cottonwoods, the first in this high drainage.

Fifteen minutes later our socks have filled with grass seeds, which drive their hard, stiff bristles into tender flesh. Picking at them only breaks off the bristles and pulls threads from cotton weave. Beneath the cottonwoods, we disturb a community of cicadas, one hidden insect screaming a solo until one after another a whole deafening chorus has joined its single-note song. There is no harmony among male cicadas (to my ears anyway), just competitive monotone.

The trees, I notice, aren't the venerable kind found along other parts of the river, like the middle Santa Cruz near Tumacácori. These are smaller and thinner and cluster closer together, their canopies locking together overhead. Here, the trees mark three channels that join at a central point before continuing on, the arrangement appearing like a giant heron's foot. Along the first phalange of the right toe, a depression of short grasses carpets what looks like a marsh, only absent of water. Flowering goldeneye ring the depression, and an occasional milkweed lifts its now ruptured pods into the wind. Where Karen and I sit together in the warmth-scented grass, grasshoppers ratchet, flies buzz, sparrows cheet.

I think about leaving something here, some personal trinket that water will carry from this beginning, bearing it through the unmovable landforms of this desert I cherish, until it comes to some quiet, sedimentary end. I imagine that this river might represent my life's course. Although I have little care to repeat my youthful days, relive my mistakes, and experience again the path appointed to me by gravity and the indifference of geology, part of me deeply believes what Socrates said about the unexamined life. I want a life worth living. This isn't about picking at the scabs of my failures. It's about understanding the reasons for the wounds. I decide that I won't leave something at these headwaters. Instead, I will take something with me. A desire, perhaps. To follow the river. To find in its bold and shining meanders a sense of redemption.

From the headwaters of the Santa Cruz River, we head south toward Mexico through the San Rafael Valley as John Russell Bartlett did in September 1851: amazed by grass. Commenting on the "luxuriant

Headwaters of the Santa Cruz

herbage," the U.S. boundary commissioner with his convoy of soldiers, wagons, and herds of sheep and cattle, rode into the northern valley from Canelo Pass to map the border between the United States and Mexico after the 1848 Treaty of Guadalupe Hidalgo, a "peace" treaty that would cost Mexico more than half of its country.

Karen and I see the same country that Bartlett's eyes fell upon: a broad landscape of tumbling hills, creased and folded upon each other, then smoothed over with a furring of sallow grasses. An emerald line of cottonwoods traces the southern course of the Santa Cruz River the whole length of the sun-blanched valley, the only other green being the scattered punctuations of oak. The scene rolling before us, one of the last remnants of southwestern short-grass prairie, is both stark and beautiful, a combination that dilates pupils and scorches retinas.

From its blurred eastern margin in Saskatchewan, the high, yellow veldt tilts toward its western edge in the foothills of the Rockies south through Arizona and into the San Rafael Valley. Buffalo grass and blue grama sprout knee high on less than ten inches of precipitation annually. These are the drought-tolerant "sod grasses," shallow-rooted and rhizome-tangled mats that once made a sturdy building material for the Laura Ingallses of the western plains.

Like characters in the *Little House* books, Karen and her younger siblings raised their own chickens, ducks, and pigs while their mother handled the household chores. Karen's father, a tall, taciturn, former navy sailor, worked two jobs to build their home, raising a two-story house out of the desert hardpan literally over the heads of his family. Then, Karen's family seemed to me to be a modern desert version of *Little House on the Prairie*; and it's no surprise that even today they still cherish the Wilder books. This is the family that I married into, simple Christian people of traditional values.

When I first began dating Karen, gaining entrance into this family was a challenge, at least where her father was concerned. Pleasing Karen's mother, as it turned out, required much less effort. But John respected hard work—it seemed there was always a remodeling project or a trench that had to be dug or a septic tank that needed excavation and scouring. John approved of calluses and sweat—and more. He appreciated people he could show how to backfill a leaching field. (Karen now thinks that bringing home willing and able boys is the only reason her father tolerated daughters!)

Karen was certainly worth the hard physical labor with her father wielding power tools, even if there wouldn't be cold beer at the end of the day. Fortunately, her mother's standard for a proper daughter's boyfriend was not as high. At the time I worked as a counselor for a summer youth camp, spending my days hiking and camping and directing a backpacking program. Kathy noticed only my best qualities. Among her daughter's boyfriends, I became "the one with the legs."

Karen would argue with her mother that all her boyfriends had legs, but she has learned from her father's example that when something malfunctions, you fix it. Simple. This basic way of living has saved our marriage—saved my life—many times. Karen has her own ideas about how marriage works, and they are anything but traditional since her prime tenet is that she has control. She stands in her commitment to me as if it were an unbreakable promise she made to the universe. It's like she's a character in one of her fantasy novels, raising a standard above a narrow, crumbling, chasm-spanning bridge, and shouting to the demon, "You shall not pass!"

Unfortunately, this kind of courage usually pulls you into the abyss. Despite my failings, which are many, she says she has always seen a glimmer of something salvageable in me. Today, after more than twenty-five years together, she still lights watch fires around our marriage.

"No Trespassing" signs mark most of the roads that branch away from the main dirt track I'm following down the valley. Another sign we see: "Grazing Prevents Blazing." Along with their independence, ranchers here prefer their privacy, and, since much of the San Rafael is fenced private land, they get it. The drivers of the few trucks I've passed, our tires throwing commingling plumes of yellow dust, eye me suspiciously. *What are city people doing on my land?* I imagine them thinking. I feel exposed, wrong. The crime rate here must be zero.

Despite the posted gates and fences, the windmills and dots of cattle, my mind keeps turning the isolated smudges of oak into dark lumbering mammoths grazing among Pleistocene hills. I've heard that the valley hasn't changed much in twelve thousand years, and I believe it. These same grasses, once chewed upon by Ice Age mammoths and bison, horses and camels, now sustain European ungulates. Ranching seems only a current phase in the natural history of this place. Although no one would argue that native wildlife haven't suffered due to human impacts, including the earliest—some believe that Clovis hunters are responsible for the extinction of the megafauna—at least one species may be adapting to ranching.

The Sonoran tiger salamander, a subspecies of tiger salamander discovered here in a stock tank in 1949 and known only from the San Rafael Valley, survives today because it takes advantage of these artificial ponds. The zebra-skinned amphibian probably once waddled between cienegas and other marshy areas prior to human settlement. Now, however, as the region has lost many of these wetlands, it prefers cow tanks, all of them within a nineteen-mile radius of the town of Lochiel, Arizona.

After spring unlocks the ponds of ice, mature salamanders seek one another out. But unlike their marshland ancestors, some of these adults don't travel overland. They can't, as least the ones with legs but no lungs. In response to a drying landscape, some tiger salamanders can keep their youthful affection for a watery world: a headful of feathery gills and membranous tail fins. Yet, their sexual organs having outgrown the rest of their bodies, these larvae—some call them "mudpuppies" or "waterdogs"—can now spawn, courting each other in the normal way with much tail wagging and pushing and nudging.

There is no amplexus among salamanders, that full-body embrace of other amphibians (frogs and toads) that comes as close to sex as any animal gets without having the usual equipment. Both sexes have a cloaca, the all-purpose organ of reptiles and birds for those most intimate bodily functions—including ones not requiring a partner. The male salamander wraps his germinal gift in a gelatinous package and presents it to his mate for her acceptance—no embracing, not even the "cloacal quickie" of the avian kind. After taking in his spermatophore, she will fertilize her eggs as they move down her cloaca, cementing them singly or in masses to stones and plants at the bottom of the pond.

The Arizona Game and Fish Department has been mapping the salamander's range in the valley to help develop a recovery plan for the amphibians. Bullfrogs, invasive and aggressive amphibians, threaten the salamanders because the frogs compete with them for food. And, if these salamanders are anything like the one our daughters keep in their bedroom, they must find the competition stiff, even overwhelming.

"Sally," an adult tiger salamander our daughters raised from a bait-shop waterdog, must believe her food drops from the heavens like manna. Karen and I really wonder about the amphibian. The only effort she makes to capture the earthworms and mealworms the girls dangle with tweezers in front of her is to slightly raise her perpetually smiling face and snap at them. Once. Most of the times she misses completely, at which point feeding time is over. Sometimes at night I walk past her glass terrarium and there she is, head lifted, staring at the ceiling as if praying.

Along with "Pin" and "Princess," two golden-eyed spadefoots (toads) the girls raised from tadpoles, Sally the salamander provides hours of entertainment. "This is what children do who don't have television," Karen explained to me the day I found them covered with mud in the yard. Jessica, Kasondra, and Melissa had created an earthen castle, complete with hidden tunnels, rivers, and a moat they had filled with water from the garden hose. Plastic toy knights and dinosaurs stood on the ramparts and along the waterways, defending the castle from a rampaging Sally-Grendel who would rise from the waters and trample her victims. Where was Beowulf when you needed him?!

Maybe instead of setting up billboards that read "Grazing Prevents Blazing" the ranching community should adopt the doggedly rugged and muscular Sonoran tiger salamander as its mascot—it's certainly a better symbol of the benefits of raising cattle in southeastern Arizona. Especially since "blazing" is now considered environmentally necessary for grassland health.

Because the San Rafael has remained largely undeveloped, these headwaters still harbor important wildlife, both diminutive—like salamanders—and grand. The Mexican gray wolf, once fairly common before our federal predator control program extirpated it from Arizona by the 1950s, may be returning here on its own. (The first Mexican wolf killed in Arizona in this federal program was trapped and shot in the San Rafael Valley in 1917.) Reintroduction efforts are under way in eastern Arizona, but there have been persistent and as yet unconfirmed reports of wolves crossing from Mexico into the San Rafael Valley. Although Tim Snow, nongame specialist with the Game and Fish, thinks these wolves are probably hybrids, he does admit that at some point wolves dispersing from the release area in eastern Arizona will reclaim this valley, "but I don't think we are there yet," he says.

Whether or not the howls of wolves curl through this valley, jaguars do occasionally pad through on silent feet, coughing that short, guttural snarl. In October of 2002, a black jaguar crossed into the Huachuca Mountains on the eastern rim of the valley. Another was spotted in the Patagonias to the west. Although these sightings are unconfirmed, in November 1971, two boys hunting ducks on the Santa Cruz River at Kino Springs shot and killed a large male jaguar, whose stomach was "full of frogs." Between 1996 and 2004, witnesses have photographed jaguars in Arizona on eight occasions. Two of these animals were caught on camera just recently using motion-sensing cameras set near the border. Over the past century and a half, credible sources have recorded eighty-four jaguars in our state, some as far north as the White Mountains and Grand Canyon. Historically, the animals ranged across the Southwest's deserts, grasslands, and forests from Texas to California as part of a territory that once may have included all of the United States before human settlement pressures reduced the cat's territory to the present range of Mexico south to Argentina. A single male cat may travel up to five hundred miles, particularly ones in search of new territory and females. This may be the case with the Arizona jaguars, the cats dispersing out of the Sierra Madre in Mexico.

Heavy nodding heads of sunflower picket the road and banks of the Santa Cruz River where we stop on the bridge at a gauging station less than two miles from Mexico. The U.S. Geological Survey measures the rate of flow at this point, 4,620 feet above sea level, where the Santa Cruz drains the 82.2-square-mile San Rafael Valley before the river enters Mexico. Today, however, the readings must register close to nil. There's water in the river, but I wouldn't call it "flow." Black liquid, mostly obscured by bent and matted vegetation, pools under the south side of the bridge. Girthy cottonwoods throw up high vaulted crowns all the way south to the border. A ribbon of silence and shadow runs beneath them.

I try to imagine what it was like this past July 16 when eighteen cubic feet of water per second surged by this station. If one cubic foot of water weighs over sixty pounds, that's like the volume of one GE refrigerator-freezer—automatic icemaker included—with five times the shipping weight bearing down on you every second. Then, I try to picture August 5. In what must have been a sustained regional deluge that swallowed this bridge, 256 cubic feet of water per second roared by. I've seen the normally sandy Santa Cruz riverbed in Tucson filled bank to bank with foam-capped waves of brown floodwater, whole trees, truck tires, automobiles, and the occasional suicidal kayaker fighting the torrent. It's why you often hear that more people die of drowning in the desert than of thirst.

Greene Ranch, 10 River Miles

I look west into the grassy hills, then north, then east—it's all grass, grass as far as I can see, grass that leans into the knobby hips of the mountains. On this bridge at the gauging station, Karen and I stand in the navel of an 1825 Mexican land grant known as San Rafael de la Zanja, a thirty-five-square-mile torso lying at the center of the valley. A little more than a mile directly to the west of us, one of the best-preserved brick ranch houses from Arizona's territorial period raises its magnificent French colonial architecture two stories above this sweeping and empty land.

Colin Cameron, a Pennsylvanian descended from Scottish Highlanders, whose loins produced some notable public figures in America, built this stately albeit lonely home. He came to the Arizona territory in 1882 at the age of thirty-four, a short, hawk-nosed son of a prosperous businessman and railroad magnate. Sporting a waxed handlebar mustache and tailored Philadelphia suit, he stepped off the train in Tucson and immediately began seeking to enter the lucrative cattle industry. After a year of inspecting the region's grasslands, talking to ranchers, and studying their methods, he found what he needed in another Pennsylvanian named Rollin Richardson. Richardson had begun consolidating ranches by buying out small-scale stockmen, including Dr. Alfred Green who had acquired a huge land grant from the heirs of the original grantee, Ramón Romero. The grant, called "Rancho San Rafael de la Zanja," occupied more than seventeen thousand acres of the San Rafael Valley and extended across the border into Sonora, Mexico. This, together with the "overplus" lands (surrounding property usually purchased by the grantee but of questionable title), gave Richardson control of nearly 153,000 acres of prime rangeland. Cameron wired friends and relatives about Richardson's willingness to sell, and in 1883 chartered the San Rafael Cattle Company.

Colin Cameron, who never saw ranching as a lifestyle but as a business, had learned that the only way to make ranching truly profitable was to go big. The following November, to the ridicule of the area's veteran cattlemen, Cameron arrived at Crittenden Station dressed in his fashionable eastern suit to supervise the delivery of sixty pedigreed Herefords. It wasn't his choice in clothes so much as his choice in cows. No one before had seen Herefords in Arizona territory. And no one except Cameron believed the highbred whitefaces could survive the winter. But every one of them did. Colin Cameron was unconventional, with fine and expensive tastes, and he didn't mind taking risks. If he was going to be a rancher in the West, he would accomplish it with style.

Colin convinced his younger brother Brewster to join him in his cattle venture, whereupon the two moved into a simple adobe house previously built here and began adding seven more rooms. Next, they enlarged a second ranch house to ten rooms, and sometime during that fall, Colin moved his wife, Alice, and their four children into this home. Later, Cameron started a third, even larger house with fifteen rooms and four baths, enclosed on three sides by porches with finished colonnettes, brick chimneys, and sash windows. A service wing extended from the rear corner of the house. Unfortunately, on Christmas Eve 1899, this third ranch house mysteriously burned, leaving only one chimney standing.

It seems members of the Settlers' Protective Association had retaliated against the rancher for his treatment of nearby homesteaders—"squatters" who had previously occupied his overplus lands and other smaller ranchers running cattle on his grant.

Stories vary, depending on the point of view, but what's clear is that the conflict revolved around the issue of these land grants and rights to the overplus lands. Beginning in the 1880s as the Apache wars came to an end, the Southwest opened once again to settlement. Without the hostile Indian threat, miners could work their claims, homesteaders plant their fields, and ranchers graze their cattle in relative peace—or so it seemed. Some "investment" ranchers began squeezing out smaller operations by consolidating vast tracts of rangeland, buying up or otherwise obtaining titles to lands originally granted to their owners by the Spanish and then Mexican governments prior to the Treaty of Guadalupe Hildago in 1848 and the Gadsden Purchase of 1854. The treaty stipulated that the U.S. government would recognize the land grants, provided that they had been recorded in Mexico and properly marked with monuments. Needless to say, land speculators and other opportunists, both scrupulous and otherwise, recognized a potential boon out West.

This was the climate surrounding Colin Cameron. He claimed much of the San Rafael Valley under the De la Zanja land grant and the surrounding overplus lands for his cattle ranch. But homesteaders and smaller ranchers saw this differently. Violence was common. One article from the *Arizona Republican* newspaper describes an assault on Cameron near Lochiel by two men who "dismounted from their horses and pulling Mr. Cameron from his horse beat him almost into insensibility." Homesteaders accused the rancher of hiring men to burn their homes and fill in their wells. Although nothing ever came of it, the governor of Sonora issued an arrest warrant for Cameron for allegedly burning the homes of two Missouri families who had settled on his Mexico portion of the grant. In the end, the U.S. government recognized the original grant of 17,324 acres, about a tenth of the land Cameron claimed to have purchased.

Karen and I continue down the road toward the town of Lochiel, and soon, where the road turns sharply south, we see Cameron's stark ranch house behind us rising out of the grassland. Dark brick walls of fired native clay raise a truncated hipped roof that extends over a wraparound veranda and supporting colonnettes. Four dormer windows and four chimneys with double cornices pierce the roof. The house reminds me of Old Main on the University of Arizona campus.

After the Christmas fire, Colin Cameron moved his family back into his second house while he worked on a fourth construction, the masterpiece Karen and I are looking at today. He based this home on a Louisiana country cottage style resembling the mid-eighteenth-century French colonial plantation house of Parlange in Pointe Coupee Parish. The two-story house has thirty rooms, including ten bedrooms (some with fireplaces), a living room, a sunroom and parlor, and a raised basement that originally contained a root cellar, coal room, and ranch store, and provided living quarters for two Chinese servants.

From these plush ranch headquarters, Colin Cameron raised his prize Herefords, the largest registered herd west of the Mississippi. He exhibited them at livestock shows in Kansas City and became a recognized authority on the breed. He was a prominent member of the National Livestock Association and chairman of the Arizona Cattle Sanitary Board, and he worked hard to improve the industry through the drafting of new stock laws. As one of his last investments before he retired from ranching in 1904, he helped form the Arizona Cattle Growers Association. In little more than twenty years, this fancy-dressed easterner with the shiny cane had become the region's most powerful cattleman.

Colin Cameron sold his ranch and property for 1.5 million dollars in 1903 to Colonel William C. Greene and eventually moved to Tucson where he built a beautiful home, which he named Lochaber, on West Franklin Street within sight of the Santa Cruz River. He lived there until his death on March 6, 1911, at the age of sixty-one.

My map calls this place "Greene's Ranch," which I find curious since it was Cameron who struggled so desperately to make his mark in this isolated valley. Apparently, the Greenes made their mark as well. Colonel Greene, who in 1903 was still years away from starting construction on his canal on the lower Santa Cruz River, came to the San Rafael Valley with a reputation—and not only for the fortune he had made with his Cananea mine in Mexico. Six years earlier, on July 1, 1897, Greene drove his buggy to Tombstone's notorious O.K. Corral and put four bullets into a man, killing him.

William Cornel Greene first arrived in Tombstone in 1880, probably drawn to the boomtown by stories of Ed Schieffelin's famous silver strike three years earlier. Greene, born into a Quaker family in Wisconsin but raised in New York, had been prospecting with his friend George Burbank in the Bradshaw Mountains near Prescott since 1877, where he hardened his muscles polishing drills in the mines after starting west at the age of twenty. Now, twenty-seven years old with powerful shoulders on a big frame, he was ready to take advantage of whatever Tombstone had to offer. He retained his eastern disposition, which served him well: his quiet charm, intelligence, eloquent speech, and the way he always parted his thick wavy hair on the left. He was a friend to everyone (and they say he could talk anyone into anything). But the West had also toughened his character along with his muscles. Greene suffered from a quick temper.

Although prospecting in the hills east of Tombstone was largely unprofitable, another venture proved productive. On June 9, 1884, William Greene married Ella Roberts Moson, a divorcee with money and two young children and who had come to Tombstone with her brother to start a cattle ranch. Ella, convent-educated, had culture to go with her California family money (in her photograph she looks a bit out of place for Tombstone in her high collar and pinned-up tresses). But she didn't have much business sense. She allowed her husband to invest her inheritance for her, which he did, serving Ella's interests for her children by purchasing a herd of cattle for her.

The following year, Greene moved his wife and family to a homestead and farm near the community of Hereford twenty miles south of Tombstone on the banks of the San Pedro River. There, William Greene the farmer pulled water from the San Pedro River to irrigate his crops, alfalfa and beans being his specialty. The hay he delivered to Fort Huachuca, twenty-five miles away. The beans he sold to the Copper Queen store in Bisbee.

Greene, like Colin Cameron, was a man who always thought big. In a few years, this kind of thinking concerning his mines would make millions for him. But on the San Pedro River, where water flowed irregularly at best, it would only get him into trouble.

While Ella Roberts gave birth to his daughters, Ella (1887) and Eva (1890), Greene worked on the construction of eight reservoirs and the ditches that irrigated his crops. He would not waste a drop of river water, notwithstanding the feelings of those downstream from him. It

Greene Ranch

wasn't as if he was hoarding the whole river. For most of the 1890s, his relationships with other people on the San Pedro were amicable. But one man, whose own farm lay on the banks of the river only a mile and a half below Greene's farm, believed Greene had dipped from the river a bit more than his fair share.

Jim Burnett had been justice of the peace at Charleston, a riverside ore-processing town whose water supply and stamp mills served the Tombstone mines from the late 1870s until their decline at the end of the next decade. If Tombstone had a tough reputation, then Charleston was its evil twin. It was a hangout for cattle rustlers and outlaws—like the infamous Curly Bill Brocius and Johnny Ringo, and for train-robbing peace officers like Billie Stiles and Burt Alvord. And it was home to Jim Burnett, a man with the penetrating eyes, haggard beard, and audacity of a desert prophet right out of ancient Israel. Burnett didn't need to sweeten his locusts with honey. But a man of God he was not (although like many dispensers of the law, then and now, he probably claimed he was doing God's work). Aside from his farm on the San Pedro, Burnett owned a ranch and butcher shop, both of which he kept stocked by fining the locals out of their cattle. He had a reputation of surprising his suspects and bringing court into session at the end of his shotgun. Judge

Burnett didn't concern himself much with arrest warrants, courtrooms, trials, and other such tedious legal formalities.

After activity at Charleston began to wane in the late 1880s, Jim Burnett set up his justice of the peace operation at Pearce, a small mining town northeast of Tombstone, and soon began expanding his farm on the San Pedro. This seems to be about the time he and Greene began to have conflicts about water, but Burnett had also alienated others in the community when he leased his land to a cooperative of Chinese farmers, foreigners his white neighbors treated with aversion and prejudice. When his tenants said they needed more water, Burnett promised them, one way or another, they'd have it.

On a hot summer day late in June 1897, Greene's children, nine-year-old Ella and seven-year-old Eva, went swimming with their friend Katie Corcoran in the river a half mile below the dam where the girls had found a shallow pool. Western folklorist and historian C. L. Sonnichsen, author of the book *Colonel Greene and the Copper Skyrocket*, explains what followed: "There are two stories about what happened. One says that when the dam went out the girls were caught in the resulting flood. As the other tells it, the dam was blown out the night before. The flood had subsided, but it had washed out a deep hole where there had been shallow water before. Katie beat the other girls in a race to the river, jumped in first and disappeared. Ella knew Katie could not swim. She could not swim either, but she plunged in and tried to save her friend."

Eva ran for help, but it came too late. News of the drownings shocked and horrified the community; the Greenes were stricken. William's grief quickly turned to rage toward the one man he knew must be responsible. He went through the motions, offering a thousand-dollar reward for information about the dynamiting of his dam, but it seems he already knew who to blame.

Burnett had loosened his mouth too often. People had heard him boast that he intended to kill Greene at the first opportunity. One rancher later claimed that Burnett had said in true western fashion: "This country is not big enough to hold him and me. Greene and I can't both live, and I'm going to live."

Whether it was an accident of fate or an act of will that brought Greene and Burnett together on that first day of July at one o'clock in the afternoon, the truth is probably lost to history. Maybe Greene had come to Tombstone only to repair his pistol. Or maybe he knew Burnett would be in town and came gunning for him. Either way, when Greene walked out of the O.K. Corral office with his pistol in his hand and ran into Jim Burnett, the game pieces fell into place. The sudden tension must have silenced the birds.

"You blew up that dam and killed my daughter," Greene told Burnett.

"No, I didn't," Burnett answered, as Greene raised his pistol and fired three quick shots into his chest. Burnett staggered through the O.K. office, followed by Greene, and finally collapsed in front of Hartford's saloon. Greene fired a fourth bullet into him as he lay on the ground, then walked east down Allen Street and turned himself in to Tombstone's chief of police.

Greene later told a reporter that he knew without doubt that Burnett had caused the death of Ella, and "when I thought of my little girl as she put her arms around my neck on the day she was drowned, I could think of nothing but vengeance." Then he quoted Deuteronomy 32:35: "'Vengeance is mine, I will repay,' saith the Lord."

His trial began five months later. After four days of testimony, the case went to the jury. Twelve citizens deliberated for ten minutes, then announced their verdict: not guilty. Greene, who was probably not in any real trouble, never spent a day in jail. But the emotional burden of his daughter's death and the Burnett killing seems to have had other consequences, with his friendships, with his reputation. And among these, his wife. Afterward, Ella suffered a long sickness during which she returned home to Los Angeles for treatment. There, almost exactly two years after her husband's acquittal, she died from cancer.

During the time his wife was in California, Greene renewed focus on his mining business. He gained titles to promising new properties in Mexico and began consolidating his holdings. Then, in April 1899, with his first real promotional venture, he formed the Cobre Grande Copper Company, soon to become his "copper skyrocket" to fortune. This was the beginning of William Greene's spectacular rise. With one hand gripping his Cananea mines and the other dipping into New York's financial district, Greene was on his way to becoming a millionaire promoter extraordinaire and the Copper King of the Southwest.

In early 1901, now forty-seven, he married Mary Proctor, a beautiful Hispanic woman half his age, the couple eloping to Phoenix because Mary's foster mother opposed the union. She was just what he needed. That same year, he organized the Cananea Cattle Company in Mexico and the Greene Cattle Company in the United States. The following year,

Greene Gold-Silver Company joined his Mexican mining organizations. Then, in 1903, he finished piecing together his ranch holdings, settling the claims of the land grants surrounding his original Hereford homestead and at last adding the crown jewel—Colin Cameron's San Rafael Ranch. Greene would eventually control over eight hundred thousand acres on both sides of the border, appending the title "cattle baron" to his reputation as copper king. In 1904, Greene formed two more companies, mostly to augment his mines: Sierra Madre Land and Lumber and the Rio Grande, Sierra Madre, and Pacific Railroad. Everywhere, in New York and out West, people started calling him "Colonel" W. C. Greene.

In the end, the Colonel fell as he rose, trailing flames all the way. On Monday, July 31, 1911, with his Mexican mines and lumber mills bankrupt and Cananea embroiled in a bloody revolution (although his new reservoir project on the lower Santa Cruz River was gathering momentum), William Greene climbed into his buggy one last time. When something spooked the horses, he lost control of the reins and stumbled into the bottom of the buggy. The team raced wildly, smashing the buggy into a telephone pole and throwing Greene hard into a fence. He walked away from the accident, but he had broken some ribs and punctured a lung. Five days later, pneumonia finally killed him.

Mary bore six children in the years of their marriage, four of them in New York City. Florence Greene Sharp inherited the San Rafael Ranch and San Rafael Cattle Company, and her family continued its operation through most of the twentieth century.

The Nature Conservancy acquired the twenty-thousand-acre San Rafael Ranch from the Sharp family in 1998 after Florence's death, while Arizona State Parks purchased this 3,550-acre southern portion and the ranch house. For more than ninety-five years, the Sharp family has been protecting this ecologically sensitive area from development. Now, with conservation easements in place that allow continued livestock ranching, the family's partnership with these organizations will continue a legacy of commitment to the land.

Although today the ranch house is closed to the public, Arizona State Parks has plans to offer historic tours and nature walks on the property.

Lochiel, 12 River Miles

Only minutes from Greene's Ranch, we drive into cottonwood-shrouded Lochiel. In 1880, this was the site of a post office originally named

Luttrell for the owner of a nearby smelting company. Four years later, Colin and Brewster Cameron changed the name to Lochiel after their home in Scotland. Then, a boardinghouse, livery stables, butcher shop and bakery, five stores and three saloons, lined the dirt streets of the tiny hamlet, which served the needs of the area's four hundred or so residents.

Today, a few buildings remain, mostly simple residences with ancient and abandoned automobiles pierced by front-yard weeds. One home, resting on its side below a private church and cemetery, looks like a half-buried can of Tecate. Narrow windows line the west side of the Quonset hut and a single stovepipe juts out center-top. There is no welcome mat, only a sign that reads, like so many others around here, "No Trespassing."

I understand that locals like to brag that Lochiel was a favorite border crossing of Pancho Villa during his cattle rustling ventures into the states. But Lochiel has been a favorite crossing place for hundreds of years, even long before there was a border to cross. In the spring of 1539, Fray Marcos de Niza (some believe) passed through here on his way to verifying the possible discovery of the Seven Cities of Cibola. Fray Marcos was on a reconnaissance mission for Francisco Vásquez de Coronado's expedition, which, a year later, would follow the monk to his supposed "seven golden cities, the smallest of which is larger than Mexico City."

Karen and I visit the monument to Fray Marcos, a large concrete cross at roadside that rises above a small square courtyard with concrete benches and a concrete floor cluttered with grasses and the dried remains of an agave. Three steps lead to a weathered bronze plaque that reads:

By this Valley of San Rafael
Fray Marcos de Niza
Vice-commissary of the Franciscan Order
and
Delegate of the Viceroy in Mexico
Entered Arizona
The First European West of the Rockies
April 12, 1539

I don't know about the "First European West of the Rockies." Even if Fray Marcos was the first European to enter what is now Arizona, he most certainly was preceded by an African. The monk had a guide, a Moroccan named variously Estevanico, Esteban, Black Stephen, and Stephen the Moor.

Estevanico was the personal slave of Andrés Dorantes de Carranza, who, along with the famous physician-explorer Alvar Núñez Cabeza de Vaca, sailed with six hundred men on the ill-fated Narváez expedition to Florida in 1528. After a series of hurricanes, crew desertions, Indian attacks, and being stranded for months in Florida, what remained of the expedition constructed five crude boats—sewing their shirts and trousers into sails and braiding horsehair into rigging—and set out for a Spanish settlement believed to be nearby on the coast of New Spain (Mexico). All but two of the crafts capsized, eighty men surviving—the one-eyed Captain Narváez not among them—to be marooned on what de Vaca would call "Isla de Malhado," the Island of Misfortune (present-day Galveston Island). Only fifteen men lasted the winter. Of these, Estevanico, Dorantes, and another named Alonso Castillo Maldonado survived the next five years as slaves of the natives. After being reunited with Cabeza de Vaca whom they thought dead, the castaways escaped their Indian captors. They traveled into the interior of what is now Texas and northern Chihuahua and Sonora, living among several tribes as healers and learning their languages—something Estevanico became quite adept at. With tales of wealth and magnificent cities to the north, the four men finally emerged from the wild at San Miguel de Culiacán, a Spanish outpost near the Pacific Coast. It was May 1536, nearly eight years after a decimated Narváez expedition washed ashore at Galveston.

Once in Mexico City, Dorantes sold (or loaned) Estevanico to Viceroy Antonio de Mendoza, who assigned the Moor to Fray Marcos de Niza. The viceroy, hoping to find the mythical Seven Cities of Gold or even the Northwest Passage to the Orient, had asked the men to lead a new expedition into the north. Only Estevanico—who probably had little choice—agreed to go.

In February of 1539, Estevanico led the party of Fray Marcos north, reaching the Río Mayo in Sonora on March 21. Frustrated with the slow progress of the friar, Estevanico left the group to scout ahead, sending back native runners bearing wooden crosses to announce the way and the country's promise of souls in need of salvation. This was the way an African Moor, probably the first visitor from the Old World to arrive here, stepped into Arizona, although no monument makes this claim or marks the place.

Unfortunately, Estevanico would not return again. When he entered the Zuni pueblo of Hawikuh in northwest New Mexico, warriors there attacked and killed him. Some accounts say that his medicine gourd offended the people because it was trimmed with the feathers of an owl, a creature symbolizing death to the Zuni. Other accounts say that Estevanico's demands for turquoise and women set them off. Either way, it's certain that Fray Marcos, faced with a suspicious and resentful Zuni, saw his "Seven Cities" more in his mind than with his eyes. And by the time he returned to Mexico City, his imagination had gained magnificent and golden proportions.

The Coronado National Memorial stands in the southern foothills of the Huachuca Mountains only twenty miles east of the Santa Cruz River and seven miles west of the San Pedro. Its placement is a compromise, I suppose, being that the Spanish conquistador's point of entry into Arizona continues to draw controversy. Various writers and historians place Coronado's route as far east as Arizona's border with New Mexico and as far west as Organ Pipe National Monument. Some argue for the San Pedro River, and, indeed, this corridor has had several noteworthy supporters, the archaeologists Herbert Bolton and Emil Haury among them. Others, however, say Coronado could have used the well-established trade routes of Native Americans along the Santa Cruz River.

With Fray Marcos in tow, twenty-six-year-old Francisco Vásquez de Coronado and his armada of mounted and foot soldiers, craftsmen, adventurers, herds and herdsmen, and more than a thousand natives, left Mexico's west coast at Compostela on February 23, 1540. Their passage through desert and grassland churned a wide swath of turned stone and mown forage. The entire region must have felt their presence, this slow parade of humans and animals. This was to be a missionary expedition, not one of conquest, the viceroy had told Coronado. But to the Spanish, it seems, bringing Christ to the heathens also meant relieving them of the burden of their riches. And Cibola had, by Marcos's account, "a great store of gold . . . and a hill of silver."

Fray Marcos must have missed his guide Estevanico, for on this return trip he seemed to be lost most of the time. Maybe he was simply stalling the inevitable. Already, other scouting parties had carried back reports of contacts to the north with natives, but these came without mention of golden cities, only primitive stone pueblos. On July 7, 1540, Coronado rode into Hawikuh, the first of the "Seven Cities," and saw the light. "[Fray Marcos] has not told the truth in a single thing that he has said," Coronado would write, "but everything is the reverse of what he said, except the name of the city and the large stone house."

Zuni warriors had drawn a line of sacred cornmeal in the sand, calling on their spirits to help them defend their homes from the Spanish. Coronado offered gifts, but the natives attacked, charging into the very hooves of the horses to fire arrows and throw rocks at several hundred armored soldiers with steel swords, crossbows, and muskets. The Battle of Hawikuh was over in an hour, as the Zuni ran out of rocks and arrows and fled into the surrounding hills. A few Zuni lay dead. No Spaniard was killed, although Coronado, as the center of attention in his battle regalia and plumed helmet, suffered an arrow wound in the leg and a blow to the head that knocked him unconscious from his horse.

In the end, the fabled Seven Cities of Cibola offered only a few broken stones of garnet and two emeralds. The battle had been for beans. One soldier, famished from the long trek over rough terrain and with supplies running low, would record these words: "There we found something we prized more than gold or silver; namely, plentiful maize and frijoles."

I smell beans cooking over mesquite fire as Karen and I drive through Lochiel. In the shade of cottonwoods, people have gathered. Their trucks and campers and ATVs are everywhere, and blue smoke lies over the place as heavy as the voices of playing children. From my driver's side window, I ask a man in a cowboy hat about the event. He says with a Spanish accent that this happens every year here. "It is the Delarosa family reunion."

The family is huge; it must have members on both sides of the fence. I drive on, following the road, and pass a man leaning on the steel mesh talking with his Mexican neighbors on the other side. Then I see the hole in the fence, the chain link pulled apart into a wide grinning mouth as if mocking the government authorities and their laws. This is a community. The border here has become as permeable as the back fence of an urban schoolyard. A gray sheet of metal, which matches the shape of the hole perfectly, hinges against the fence like a door. It is a door swung open. A garden hose links one house on the American side to some shrubs on the Mexican side, and I imagine Lochiel was once a larger town until someone—maybe John Bartlett himself—threw a line on the ground. Here, for now, the border seems only hypothetical, an abstraction of mapmakers and politicians. The plants and animals, even the people, pay it little regard. The only concern is which cottonwood offers the most shade.

I'm searching for cottonwoods for another purpose, however. I'm looking for the path the river takes south into Mexico. Karen and I drive from Lochiel, heading north again, but then I see a brace of tracks quartering away toward the southeast, toward the Santa Cruz. My map shows a road to the river, although this doesn't seem to be the one. This one slashes through the grasses and contours the hills in the general direction of the river but then draws up short at a gate and a fence. I could open the gate and drive through if I wanted to—though a warning sign reads:

ALL PERSONS AND VEHICLES MUST ENTER THE UNITED STATES AT A DESIGNATED PORT OF ENTRY ONLY. THIS IS NOT A DESIGNATED PORT OF ENTRY. ANY PERSON OR VEHICLE ENTERING AT THIS POINT IS IN VIOLATION OF 19 USC 1459 AND/OR 19 USC 1433 AND IS SUBJECT TO A $5,000.00 PENALTY.

VEHICLES AND MERCHANDISE CONTAINED THEREIN ARE SUBJECT TO SEIZURE

U.S. CUSTOMS SERVICE

The words are repeated in Spanish. The sign, however, faces only north. If I were driving across this point from the south, I'd probably not bother to read it. I *couldn't* read it until I had already crossed over. The irony makes me smile. Fray Marcos, Coronado, and the countless denizens who have followed them into this rich land, haven't and won't ever bother with "designated ports of entry."

I stop the car and get out to look south into Mexico. Leaf-heavy cottonwoods crease the hills only a quarter mile to the east. My shoulders are burned, my reddened skin approaching the color of my shirt. This will be the most unfamiliar part of my journey, although I know the direction I must take. I can see it laid out to me before my eyes. Beyond the fence line, the boundary, the white monuments erected between peoples. Beyond the warning sign to travelers. Beyond the soon-coming border fence and its steel barricade to vehicles. The place where the valley folds on itself and the river pays no attention to monuments or warnings or lines on maps.

San Rafael Valley at the Arizona border

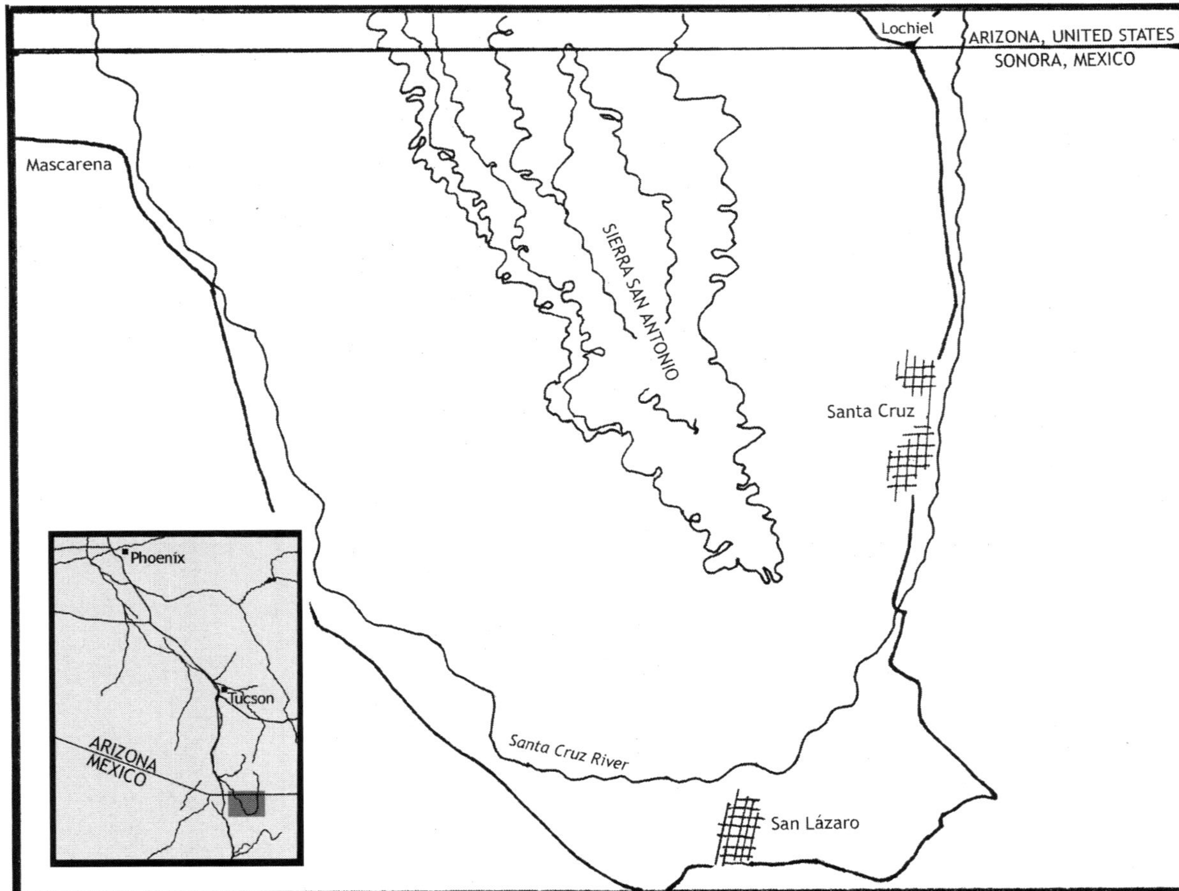

The Sonoran Reach

3

Mexican Water

The Sonoran Reach

Only redemption can reunite an exiled soul with its root. The holy person, however, can hasten redemption and help mend heaven and earth.

—Annie Dillard, *For the Time Being*

Santa Cruz: April, 19 River Miles

"I know the place where you can cross into Mexico," Alan Weisman told me a few years ago when I described to him my intentions of following the river across the border. "And when you cross, go in the spring. You have to see the orchards of Santa Cruz in the spring."

Later, I looked up Alan's book *La Frontera*, in which he writes that outside the village the river's bottomland is planted entirely in apple trees: "In Santa Cruz, which means Holy Cross, Easter arrives perfumed with apple blossoms."

There's only one way to enter Mexico—through a fence. And, although this will soon change, all I have to do today is open the gate and walk through, following a dirt road that's been here longer than this boundary and the six strands of barbed wire that mark it. The road heads directly south, paralleling a cottonwood-hemmed Santa Cruz River on my left. I'm seven miles from the Mexican village of Santa Cruz.

The valley before me dips and ripples southward, low tawny hills on my left, splashed with olive as if the trees were flicked from a watercolorist's paintbrush. To my right, the smooth mounds of the Sierra San Antonio tumble out of the Patagonias and spill across the Mexican border to lie inside the fishhook of the Santa Cruz River.

The scene is the same today as the one John Russell Bartlett would have witnessed 155 years ago, a valley "covered with the most luxuriant herbage, and thickly studded with live oaks; not like a forest, but rather resembling a cultivated park."

Bartlett and his entourage were headed for Santa Cruz on a mission that had little to do with his border survey but was probably one of the noblest acts of his otherwise inept career as U.S. boundary commissioner: the return of a rescued Apache captive named Inez Gonzales.

On September 30, 1850, fifteen-year-old Inez Gonzales, her aunt and uncle, another young girl and her brother, Francisco Pascheco, were being escorted by ten soldiers to the town of Magdalena to celebrate the annual feast of San Francisco. Two days out from Santa Cruz, a band of Apaches attacked them, killing eight of the soldiers and Inez's uncle and carrying off her aunt and the three children. The captives spent the next several months working as slaves and living off baked agave hearts until the Apaches sold her aunt and friends to traders. Inez never saw them again. The Apaches eventually sold Inez to another group of traders, who were on their way to Santa Fe with her when Bartlett encountered them and demanded her release.

Three months later and almost exactly a year after her capture, Inez returned home. On that day, a few miles north of Santa Cruz, a large group of people dressed in gaudy holiday costumes approached Bartlett's party. "As we drew nearer," Bartlett writes, "Mr. Cremony helped Inez from her saddle, when in perfect ecstasy she rushed to her mother's arms." In *Life among the Apache*, John C. Cremony recalls that "Mr. Bartlett conceded to me the privilege of placing Inez into the longing arms of her mother, who, after repeated embraces, and amidst alternate tears, prayers, thanksgivings and joyous cries, yielded her place to the strong but inferior claims of other relatives and friends, all of whom ardently and most affectionately embraced her by turns." Cremony would go on to say that the little reunion scene on the banks of the Santa Cruz River was the most inspirational he'd ever witnessed and formed a "delicious oasis" among all his years of seeing the horrors of what people could do to one another.

Enormous cottonwoods rise out of the riverbanks and erupt into the sky like glaucous-plumed thunderheads. Male vermilion flycatchers pump their dark and undersized wings to corkscrew slowly above the highest

Santa Cruz River in Mexico

branches. Everywhere is the smell of water and rock, a quiet gathering of pungency, of shining runnels and algal slack water trapped in the wallows and trackways of cattle.

I look south toward the Sierra de Pinto, whose upwelling flanks bend the river from south to north, and the distant Sierra Azul where researchers recently captured a wild ocelot on film, collected from a camera that my oldest daughter, Jessica, helped set up. Trotting along a rocky drainage only twenty-five miles from our border with Mexico, the endangered spotted cat was a pleasant surprise and quickly made news. It's been forty-five years (1964) since an ocelot was documented in Arizona, although unconfirmed sightings occasionally leak out of the Chiricahua Mountains in the southeast corner of the state. Jessica says the animals may be crossing from Mexico, perhaps even using this river as a pathway north.

My daughter, in addition to pursuing her studies at the University of Arizona, is volunteering with the Sky Island Alliance, assisting international efforts to record the movements of our border-crossing jaguars. The nonprofit conservation organization works with scientists, volunteers, and landowners to protect and restore our so-called Sky Island region. "Sky island," a term coined by Weldon Heald in his 1967 book about the Chiricahua Mountains, refers to the kind of isolated, desert-encompassed mountain unique to the Southwest. (An artifact of the Basin and Range Province, an archipelago of forty or so such mountains stretches from Arizona's Mogollon Rim to Mexico's Sierra Madre Occidental.) Like islands in a vast ocean, it's the insularity of these mountains that gives their inhabitants a kind of inertia, a "flywheel of evolution" whereby animals and plants either adapt or go extinct. As a result, our sky islands hold some of the most amazing, one-of-a-kind, and irreplaceable wildlife in the world.

Jessica has been traveling into remote country in Arizona and Mexico with Sky Island staff to set up motion-sensor cameras to document this wildlife, hiking and camping with men who spend most of their days away from modern conveniences like showers and toilets. This seems to be where her interests in nature have led her. As a college representative of the local chapter of the Wildlife Society, she organized wild game feasts, traveled to Alaska for a national conference, designed logos for T-shirts, and learned to two-step to country music. (I remember the time she called her younger sisters from a bar in Flagstaff saying, "Beer is not so bad if you drink enough of it!") Jessica, my wild child. To think that when she was nine, doctors diagnosed her with juvenile rheumatoid arthritis after she curled up into a ball and began screaming from pain. For years afterward Karen and I wondered if she would recover and live a normal life. She eventually did. Though I'm not convinced that hanging out with unshaven Ed Abbey types is normal.

Jessica's connection with Sky Island Alliance was part of the reason I asked her to join me last April for my first visit to this valley. Jessica, in turn, introduced me to another U of A wildlife biology student and Sky Island volunteer. Guillermo was born in Nogales, Sonora, and would be a helpful guide, I figured, since he knew the language and the area, with the added benefit of understanding its natural history.

But it wasn't ocelots or jaguars that interested me last April. I came to see the apple trees. In pockets along the river between the border and the tiny agricultural community of Santa Cruz, the orchards were casting a faint, sweet fragrance into the air. Gray trunks of multiple configurations twisted up out of grasses splattered with dandelions as if the flowers were canaries foraging among the blades. The wispy trees

Apple orchard in Mexico

seemed to lean toward the river, their crowns showing dark green behind white petals so diaphanous they shone through the branches as pearl.

It was just as Alan Weisman had said. At the height of the Easter season, an entire valley south of the border resurrects itself with an apple bloom.

We walked through an apple orchard to hike along the river. A cottonwood grove stretched down the far bank, but the near side was open to the sky. From somewhere, a gray hawk screamed among the branches—the raptors, like hope, always seem to be just beyond sight. Plainly visible to us, however, in its dark robe of feathers, a great blue heron slowly toed the algae-clotted river margin, the bird's huge feet leaving Triassic prints in the mud.

Although in *The Lessening Stream*, Michael Logan calls this part of the Santa Cruz "a broad and shallow creek in the service of livestock and agriculture," I saw more than cows and apple trees. Unmistakable signs showed the river's heavy use. Cattle had trampled its banks. Canals diverted its water. But, unlike other reaches of the river, particularly those near Tucson, water still flowed. Erosion hadn't downcut the river and trapped it into a steep-walled channel reinforced with gunite. Native fish, like the longfin dace, continued to send ripples across its glassy surface. Rivers will probably always be in the service of people. The question is, I guess, for Mexicans and Americans alike, will that service be sustainable?

A mile north of the village, I can see the bell tower of the church with its eggshell dome and simple cross. It is a beacon above dry desert hills, even without the sound of ringing bells, and I head straight for it. Like many Mexican towns in Sonora, the church will be the center of the community.

I walk along streets of swept concrete. Tarnished metal roofs slope over simple homes of white stuccoed adobe, the plain walls opening into a single doorway of rough boards or iron-fronted windows. Children play with a soccer ball in the street while old men in blue work shirts and stained cowboy hats sit in the shade of their porches. I smell oil and dirt, the scent of creosote released from sun-warmed railroad ties. The ubiquitous breedless dog trots by, its ribs jutting through the paper sack of its skin.

The church tower leads me to the zócalo, the town center of Santa Cruz and, as with many such Mexican plazas, a raised white gazebo with a barrel-tile roof forms its hub. On quiet evenings, I imagine the place crowded with young people, girls dressed in black skirts and heels, boys in newly washed trucks circling the square with norteño music blaring. But now, in the middle of the day, it's deserted. Only tall Italian cypresses gather on the brick plaza.

A horse-drawn road grader rests in the tall grass outside the town hall. The building's Spanish-mission façade is white and trimmed in burnt sienna. Just left of the main arch, the painted seal of Santa Cruz depicts the community's principal economies: cattle ranching and apples. But also taking a prominent place on the seal is a Yaqui deer dancer. Also a symbol of the state of Sonora, the deer-headed image recognizes the Cahita-speaking indigenous people of coastal Sinaloa and Sonora who are close cousins of the Maya. This one shakes his gourd rattles and crouches between the words "Santa Maria de Suanqui," the original village name referring to the inhabitants of the Río de Santa Maria, the first Spanish name for the Santa Cruz River. Above the picture of apples stands the church, and at the center of the seal is a golden cross, inscribed with the year 1637, the date Juan Munguia Villela came to the region and established some ranches. The village of Santa Cruz lays claim to more than 370 years of European history, while at the same time acknowledging so much more.

Last April, with the entire valley in apple bloom, Jessica, Guillermo, and I climbed these steps to speak with Bernardo Espinoza, the presidente municipal of Santa Cruz. With Guillermo translating, I asked him about the orchards, said that I wanted to visit them, and I remember how the young official with the thick pugilist's nose, serious but approachable eyes, and dark mustache suddenly jumped to his feet to take us on a private tour of them. We convinced him this wasn't necessary.

While Bernardo spoke about the value of the river for the orchards, as well as other crops like beans and corn, my eyes kept shifting to the wall behind his desk, to a framed old photograph of a man obviously posing for the camera. He wore a disheveled military dress coat from the Apache wars, the diamond and chevrons of a first sergeant's insignia clearly visible on his left sleeve. His white-gloved hands gripped the muzzle of a Springfield rifle (just like the stage prop Geronimo holds in the famous Camillus Sidney Fly photograph), the butt of which rested at his boots. Black hair fell from his wide-brimmed hat, framing a face with lips pulled tight into a grimace. His left eye seemed dead under the ironic lift of its brow, but his shining right eye looked straight at me.

Bernardo said it was an 1887 picture of Mickey Free. Manuel Rojas, the Mexican historian, had given it to him when he came to Santa Cruz to do some research on the Apaches. "Mickey Free," he said, "was a mestizo who was born here in 1847."

Stories abound about Mickey Free. His life has been fictionalized, romanticized, mythologized. Born Felix Telles, some say he was a child of mixed blood, half Apache, half Mexican, that his mother, in fact, was a companion of Inez Gonzales and his father one of her captors. Others say he was the son of a Mexican woman and an Irishman. Either way, after Apaches supposedly abducted him at the age of thirteen, he started down the road to becoming a feared warrior and notorious army scout. People said he could track a shadow on a rainy night. Tom Horn, the infamous Pinkerton detective and assassin who knew Mickey Free, described him as having fiery red hair and a face like the map of Ireland, saying, "I dared not send Mickey in charge of my prisoners, for he would have killed them all."

But the British historian Allan Radbourne has a new take on the story. In his 2005 book, *Mickey Free: Apache Captive, Interpreter, and Indian Scout*, Radbourne strips away the mythology and adds flesh and blood to the boy stolen by Apaches. With three decades of research to back him up, Radbourne claims that Felix Telles wasn't the "half-breed" that people called him but a Mexican, the illegitimate child of two teenage lovers. Young Felix apparently did have auburn hair, but it seems likely to me that people described Mickey Free as a fiery-headed, "half-Irish, half-Mexican, half-Apache, and whole son-of-a-bitch" because they knew the man they thought was his father.

His name was John A. Ward. He came to America from Ireland in the 1840s or '50s and eventually arrived at Tubac in 1857 and began working as a lumberman, skidding logs out of the Santa Rita Mountains for markets in Tucson. By 1859, Ward had settled onto a 160-acre ranch among the cottonwoods and willows of Sonoita Creek twenty miles northwest of here and built a comfortable adobe home. It was to this place that John Ward brought from Santa Cruz a twenty-eight-year-old single mother, Jesús María Martínez, who became Ward's common-law wife. Felix Telles was one of her children.

Then, on January 27, 1861, while Ward was traveling on business in Sonora, nine Apaches attacked his home. Fortunately, the surprise arrival of two Americans thwarted the raiders' intentions of capturing Jesús María and her children, including the couple's ten-month-old daughter. Felix, however, was not at the house but out in the fields tending to Ward's sheep and goats. A second raiding party, chasing after the livestock, discovered the boy hiding in a peach tree. The Apaches told him to come down, and he did.

Within a month of the kidnapping, the United States responded by bringing open warfare to the Apaches. Second Lieutenant George Nicholas Bascom, believing that Chiricahua Apaches were involved, blamed Cochise, the Chiricahua chief, who denied responsibility and insisted the culprits were White Mountain Apaches. In what has become known as the "Bascom Affair," hostages were taken, stages ambushed, wagon trains burned, and people on both sides killed. Chiricahua Apaches wounded John Ward and one of Bascom's sergeants. No one, however, learned the whereabouts of Felix Telles.

Eleven years later, on November 29, 1872, Brigadier General George Crook rode his mule into Camp Apache, an army post on the White River in the heart of the White Mountains. Crook was preparing to launch a campaign against Apaches who refused to remain on the reservation, and he had decided to enlist Apaches as scouts. The quiet, taciturn officer with the sandy beard and piercing gray eyes believed, unlike some of his contemporaries, that only Apaches could track down Apaches. Over the next few days, Crook's second lieutenant, John Gregory Bourke, recruited forty-seven young men. Among them stood a lean, 135-pound, fair-skinned redhead, whose appearance the soldiers thought embodied a character from a popular and funny Victorian military novel by Charles Lever titled *Charles O'Malley: The Irish Dragoon.* The soldiers dubbed the redhead "Mickey Free."

The name, whether spoken in respect or spitted out in disgust, soon gained a reputation. In the Battle of Big Dry Wash on the Little Colorado River, First Sergeant Mickey Free and his scouts of Company B killed the Cibicue Apache chief Nantiotish and half of his men. Mickey Free tracked renegades like the notorious Victorio and Chihuahua. He single-handedly captured an escaped Tonto Apache, the brother of the tribe's chief, riding into Fort Apache with "Josh" slung over his saddle, "not dead," Radbourne writes, "but dead drunk." Even Geronimo respected the Apache scout, telling General Nelson Miles after surrendering in 1886 that he had left the reservation because Mickey Free had plotted to kill him.

After his discharge from the army in 1893, Mickey Free spent the remainder of his life among the White Mountain Apache until his death in 1914, outliving four wives and producing four children, two of whom grew to adulthood.

Bernardo removed the photo from his office wall and handed it to Jessica. Then he gave her some pages describing Mickey Free's life. We looked at a picture of a man confident about who he is.

Since learning about Mickey Free after first seeing his photograph in Santa Cruz, I've begun to doubt the "kidnapping" part of the story, at least the part concerning the Apaches. Mickey Free lived fifty-three years with the Apache and had plenty of opportunity to return to his own people. At one point, his half brother, Santiago Ward, even visited with him and invited him to come home. He refused.

And why wouldn't he? As a boy, isn't it likely that Felix became enamored with the idea of the Apache, fearless and free? He certainly had heard the stories of the nomadic bands of warriors, living on the land as if there were no other place. When I consider his young life, a Mexican teenager taken from his homeland to watch over someone's sheep and goats in a foreign country, it's easy for me to believe that Felix wanted something else for himself. Perhaps, as the Chickasaw poet Linda Hogan says, he only wanted to explore "the world of waters and the universe of life that dwells above our thin roofs."

Felix wasn't hiding in that peach tree. In January, the tree would have been a leafless tangle of sticks, a ship's mast with its crow's nest. He didn't climb the tree to hide from the Apaches. He climbed the tree to get their attention.

Outside the Public Information Building, the church stands on my right, its bell tower lifting narrow arches until each is penetrated by blue sky. Cracks in its rose-colored stucco give an appearance of age, but I remember that Bernardo had said the structure was only a hundred years old. I step up five concrete risers and push through a pair of huge wooden doors.

Santa Cruz's original Jesuit mission and church, founded by Father Eusebio Francisco Kino in 1693, is long gone, and I doubt that there's much evidence where it originally stood, since Apaches burned it to the ground in the fall of 1768 and it was probably never rebuilt. Kino named it "Santa María de Suamca," combining in his usual way both the Spanish and native tongues (*suamca* is an O'odham word meaning "immaculate"). Suamca was only one of the more than two dozen missions the tall, dark-skinned Jesuit with the penetrating eyes and pronounced brow would establish during his years in northern Sonora. His was a twenty-year presence that would do more than change maps, blaze new trails, and raise gleaming mud-walled temples on the "rim of Christendom."

He was born Eusebio Chini, the only son of Francesco Chini and Margherita Lucchi, in 1645 in the small northern Italian town of Segno. (Because of his name's confusion with "Chino," which in Spanish means "Chinaman," he later changed the spelling to "Kino," possibly because Spanish officials of his time were already highly suspicious of foreigners and Kino preferred not to raise any eyebrows concerning his origins as he traveled in New Spain.) Growing up in the province of Trent in the border region of Italy and Austria, the young Kino worked his parents' farm before leaving for Innsbruck, Austria, to study rhetoric and logic at the college of Hall. In college he showed a propensity for mathematics, but it would be his skills with language and even farming that would be

Church at Santa Cruz in Sonora, Mexico

most helpful to him, for it seemed that God had placed a calling upon his life—Kino only needed to realize it.

While at college in Innsbruck, he fell ill and came close to death. He prayed to Saint Francis Xavier for his recovery, vowing that he would dedicate himself to God and become a Jesuit if he lived. He kept the promise. In 1667 at the age of twenty-two, Kino pronounced his first vows. For another decade he studied mathematics, philosophy, geography, cartography, and astronomy at the Jesuit college in Ingolstadt, Bavaria (converting one of the school's towers into an observatory for his astronomy club). On June 12, 1677, at Eistady, Austria, the church ordained Father Eusebio Francisco Kino as a Jesuit priest. Soon he would leave his native country forever.

Kino arrived at Vera Cruz in 1681 after some misadventures of biblical proportion. First, he missed the fleet bound for New Spain from Cádiz by one day when he and eighteen other missionaries encountered bad weather and became lost during their passage across the Mediterranean. It would be two years before he could board another vessel. Unfortunately, this one would never make it out of the Bay of Cádiz, foundering in the bay's mouth in a storm that nearly destroyed the ship. Kino must have felt like the Apostle Paul after he and his companions were dragged from the sea and returned to shore with not much more than their skins intact. Kino waited another six months for the next ship.

From Vera Cruz, Father Kino traveled to the Jesuit mission at Mexico City. (Several historians believe it was at this time that he changed his name from Chini to the German spelling of Kino, but it may have also been during his years in Spain.) After demonstrating his spirit and fortitude during two failed attempts at establishing missions in Baja California, Kino, in 1687, came to the then unknown and inhospitable northern territory known as Pimería Alta, the Land of the Upper Pima Indians. There, on the Río San Miguel at the native village of Cosari (about fifty miles south of here), he founded his first mission: Nuestra Señora de los Dolores, Our Lady of Sorrows. The church he began building that same year became his new home and would remain so until the end of his life.

Santa María de Suamca was Father Kino's eighteenth mission, his fifth on the Santa Cruz River, which he established a year or so after making his way farther north along the river valley and founding his more famous missions at Tumacácori and San Xavier del Bac. Suamca was the *cabecera*, the church headquarters, for the nearby missions, called *visitas*, at San Lázaro and San Luis Bacoancos.

Of the handful of missionaries who lived at Santa María Suamca during its seventy-five-year life, one of the more colorful was Father Ignacio Xavier Keller. The tall, blond Moravian arrived on April 20, 1732, to what he would later describe as "scattered and uncivilized Indians" who "did not know how to plow." Unable to get provisions or build a church or even an adobe house for himself, Keller "persevered living for years in a straw thatched hut like the natives, sustaining myself and them on the alms I would go out and beg for."

Padre Keller was a hardheaded, obstinate loner, preferring to do things his own way, which didn't always sit well with the Spanish authorities. Some even claimed he was abusive of the natives and responsible for the Pima Indian Revolt in November of 1751, a charge Keller had to defend himself against before the viceroy in Mexico City. But the people of Suamca insisted on the return of "our Father Keller," and he was sent back to the mission somewhat exonerated.

For nearly three decades Keller served the native population (reputedly baptizing two thousand converts in a single day) who inhabited the ten thousand square miles of Pimería Alta. He died in mid-August 1759, while giving the last rites to one of his charges.

Father Barrera, a thirty-three-year-old Spanish *criollo* born in Puebla, Mexico, replaced Keller at Suamca. But he would be the last Jesuit appointed to the mission.

In the summer of 1767, Spanish soldiers arrested all the Jesuits in Pimería Alta and sent them to the Port of Guaymas. There, they shared a boat bound for San Blas to be force-marched through the coastal jungles of Nayarit and Jalisco on their way to Vera Cruz and then Spain. Father Barrera spent his last ten years imprisoned at Córdoba, Spain, where he died at the age of fifty-three.

It's interesting to me that King Charles III of Spain, in typical political fashion, only vaguely explained his reasons for ridding his lands of the black-robed priests, saying he was "moved by weighty reasons, conscious of my duty to uphold obedience, tranquility, and justice among my people, and for other urgent, just, and compelling causes, which I am locking away in my royal breast."

The summer following the Jesuit expulsion, Barrera's Franciscan replacement, Father Francisco Roche, arrived at Suamca. His assignment there, however, was short-lived. That November, Apaches attacked the village, running off with oxen and cattle. Then they returned and, as another priest later described, "burned the church and the houses of the mission and put everything to the sword and flames." The Apaches gave

Cemetery at Santa Cruz in Sonora, Mexico

leave to Father Roche to gather together the few survivors and flee south to the mission of Cocóspera.

Eventually, however, the Spanish would return to Suamca. After a short and troubled beginning in 1776 at the San Pedro River, where Apaches killed more than eighty people, the Spanish presidio of Santa Cruz de Terrenate moved here in 1787. It was the garrison's fourth and last move, and it would remain until 1821 and the end of Spanish rule. It's possible that the Spanish built the wandering presidio on the same site as the abandoned mission of Suamca. What's certain, however, is that the town and its river would no longer keep the name Father Kino gave them. Santa María de Suamca would become Santa Cruz, and the Río de Santa María would soon be known only as the Santa Cruz River.

San Lázaro, 26 River Miles

South of Santa Cruz, I wade in the shallows wearing shorts and Tevas. River walking, feeling the peristalsis of the water's current against my body, is a baptism I return to often. In fact, I was *actually* baptized in a desert river, dunked by a Baptist minister into the Salt River, a tributary of the Gila like the Santa Cruz, with all the ceremony of John the Baptist at the river Jordan. My parents had raised me Catholic, but in my twenties I felt the need to reconfirm (or recon*form*) my faith in a way more Protestant, accepting a heritage handed down to me as a foundation and then building on it something new, more personal. Naively, I would leave behind the religion of God and go looking for God himself.

Annie Dillard says that in creating this world God did not extend himself but withdrew himself, humbling and obliterating himself, leaving outside of himself this domain. If it's true, as I believed, that he does not intervene in this world except rarely, then I had to find a way to stir God. To get his attention. This was a dangerous prospect, I would learn. Job might have counseled me that it's better to be ignored by God. Certainly Job's wife would have. There's a certain trepidation that comes when the Creator of the universe suddenly turns his gaze upon you and says, "Oh, there you are."

My first Holy Communion and altar-boy training, catechism and confirmation, even my stepfather who directed the choir at Saint Odelia's and entertained dour Catholic monsignors at our home in Tucson, couldn't have prepared me for my encounter with God. To my Catholic friends and family, I became a "Jesus freak" when, in the cellar of an ancient Baptist church, I committed my life, such as it was, to Christ. It would be years before I understood what the commitment meant, and that God doesn't take such promises lightly.

Like the prophets and holy men of old, I, too, found God in a barren wilderness, although I hadn't initially gone looking for him there. Mine was more of a forced wilderness experience, which followed some brilliantly destructive choices on my part. But God was there, waiting for me, and, though I was no saint, he had work for me to do. To reach the light, the seed must empty itself of everything but faith and grow in darkness; and even in the light, photosynthesis must have a dark reaction.

My friend Fenton Johnson, describing his own passage "through the briars and thistles of faith and desire," says that one "embarks on an interior journey at peril of one's whole being." This I discovered to be true. As the third-century patriarch of hermit-monks, Saint Antony of Egypt, discovered, in this wilderness you're just as likely to meet the devil as you are God. Maybe it doesn't make a difference. Encountering either one will transform you.

With the mountains named for the austere desert monk behind me, I enter the town named for the man Christ raised from the grave, San Lázaro. Here, the Santa Cruz River renews itself, beginning its trademark U-turn to loop around the south-pointing finger of the Sierra San Antonio before flowing northwest toward Arizona. I head for the town plaza, walking along divided streets of packed dirt. Pickup trucks pass by and from somewhere I hear music, the traditional norteño style with its strong rhythm drawn from the twelve-string bajo sexto guitar and accordion. My European ears always hear polka, the music's Bohemian roots, and my mind says "chicken scratch," the dance style of the Tohono O'odham near my home.

Giant eucalyptus trees tower over the plaza, whose white outer wall and benches enclose the typical tiled gazebo. Adjacent to the plaza stands a simple collection of buildings, a health clinic and the Centro Comunitario para la Conservación del Río Santa Cruz, or CCC-SC, my destination today.

San Lázaro, like the village of Santa Cruz, is a ranching and agricultural community, whose nine hundred residents depend on the perennial flow of the river. One way that the river's importance to the community has manifested itself is in a wonderful conservation program called Los Halcones, the Falcons, which operates out of the center.

Los Halcones rose out of the inspiration of Joaquin Murrieta, the culture and conservation director of the Sonoran Institute's Sonoran Desert Ecoregion Program. The Tucson-based Sonoran Institute works to help communities in the West, "through civil dialogue, collaboration, and applied knowledge," make decisions and policies that respect both the land and its people. Joaquin, who grew up in northern Sonora and received his PhD in renewable natural resources at the University of Arizona, thinks that one way to help preserve the Santa Cruz is to engage people on both sides of the border whose lives depend on the river. At San Lázaro, Joaquin began meeting with local ranchers and farmers, but it soon became apparent that better progress could be made by involving the town's youth. Los Halcones was born as the Sonoran Institute rehabilitated an old clinic building, giving the teens a place where they could meet and organize cleanup campaigns and field trips for monitoring the health of the river, taking water samples and surveying its bird life.

Outside the conservation center, I meet Nurse Blanca Ophelia Galaz, who works at the clinic. She unlocks the door and lets me into a large open room with wildlife and watershed posters on its walls and bookshelves crammed with field guides. She hands me a packet of materials in Spanish with the heading "El Desconocido Río Santa Cruz" (The Unknown Santa Cruz River) and containing photographs and a hand-drawn map of the region. A subheading says something about the valley being irrigated by the Santa Cruz, "whose pure, crystalline waters follow a circular course, flowing out of a cienega about a league north of Santa María Suamca before getting lost in the burning land of Arizona."

Nurse Blanca tells me about the students who are currently involved with some bird surveys, and no sooner do I figure out what she's saying when three teenagers pop in the door. Word of my visit—some strange gringo interested in the river—has traveled fast. Even through the language barrier, I can see that the teens are excited and proud of the restoration work they're doing. A boy and a girl hold hands and tell me they have identified 183 species of birds along their river. Fourteen-year-old Lalo Luzania knows their names in English and Spanish. He wants to be a guide for visiting bird-watchers.

One of the goals of Los Halcones is to work with ranchers around San Lázaro to find alternative sources of water for cattle, excluding them from the riparian area where they trample banks and graze on young cottonwood and willow seedlings that would otherwise provide food and shelter for birds and other wildlife. The cattle also pollute the river, which the people of San Lázaro draw upon for drinking water.

They have their work cut out for them. All the same, Los Halcones encourages me. Following after the town's namesake, these kids just might resurrect a river.

While Los Halcones may be the hope of one reach of the Santa Cruz River, a rancher in Mexico has already been striving for years to preserve an entire watershed. Carlos Robles owns Rancho El Aribabi, a ten-thousand-acre spread that reaches from the Río Cocóspera into the Sierra Azul in northern Sonora. His ranch house rests on a saguaro-studded hillside above the river, which today is a fencerow of cottonwoods and willows that screens the corrugated eastern horizon. Ten years ago there were no trees, and the river was a stinking, muddy ditch of eroded banks and a dead wickerwork of Bermuda grass. Then Carlos removed his cows.

El Aribabi in Sonora, Mexico

I first met Carlos after Jessica and I had traveled to El Aribabi to meet a group of Sky Island Alliance researchers and volunteers who were conducting a wildlife inventory of the ranch. Our invitation had come from Sergio Avila, an SIA biologist who was born in Mexico City and grew up in Zacatecas at the southern end of the Chihuahuan Desert. With a master's degree from the Universidad de Baja California, Sergio has worked with wildlife ranging from mountain lions to sea lions. Upon our arrival, Sergio and Carlos found us at the front door of the ranch house, where the two men and Jessica greeted each other as if they were old friends. "This is my dad," Jessica said, introducing me to Carlos.

We spent our first day hiking rugged canyons, searching for mammal tracks, and changing the film in several motion-sensitive cameras. Sergio led the way, while Cynthia Wolf, a freelance biologist, wilderness outfitter, and tracker extraordinaire, scanned the sandy soil and mud for animal sign. I wondered how she could make any sense of what tracks she found with her Australian shepherd, Lash, and Sergio's hyperactive black cocker spaniel, Remi, leaving prints all over the place. (Cynthia, brushing away her blonde hair and showing her deeply bronzed shoulders, had trained Lash well. To demonstrate, she would ask her a question: "Lash, would you rather be married or be dead?" at which point Lash would roll over on her back and lie still.) Jessica kept notes for each camera, and another volunteer named Rich monitored our GPS coordinates.

In the evening, we rejoined Carlos at his ranch house and descended upon his kitchen. While Sergio reheated some carne asada he found in the refrigerator, Greg Scott and David Courtland, two volunteers from Nogales, prepared what they called "rajas de chile verde en crema," sautéing onions with roasted chilies, adding white Mennonite queso with a can of Media Crema, and wrapping the savory result into fresh tortillas. We ate the food, standing around Carlos's gas stove, warming ourselves next to the grilling tortillas. After dinner, Carlos brought out a bottle of El Jimador and poured shots and passed them around. Then, raising his glass, he spoke in a loud voice, "Al futuro del jaguar y el ocelote," offering a toast to the future of two amazing animals only recently seen on his ranch. The tequila, smooth and warm in my throat, settled into my aching joints.

The next morning, over a breakfast of huevos rancheros, Carlos talked about his business, how he had moved away from cattle ranching and had begun to diversify into less environmentally destructive activities such as hunting and ecotourism. He had turned his ranch house into a retreat center, its many bedrooms with fireplaces, its central, kiva-like living room and open porches, refitted for large or small groups of students, researchers, and sportsmen. "I can do anything people want," he told me.

"Horseback riding with a barbecue?" I asked.

"Sí. And campouts, nature and bird tours, guided hunts."

"What about a bed and breakfast so I can bring my wife?"

"Sí."

El Aribabi is one of the premier ranches in northern Mexico, hosting at least thirty endangered and/or threatened species. The ranch boasts 165 species of birds, including gorgeous, parrotlike elegant trogons and rose-throated becards. Green rat snakes and Gila monsters patrol the riverbanks where javelina and ocelots come to drink. Troops of coati, a long-snouted mammal that looks like a cross between a raccoon and a spider monkey, root in the undergrowth for grubs and tubers. And Coue's white-tailed deer strip ball mosses that fur the streamside hackberry trees.

Recently, Carlos Robles and Sky Island Alliance entered a groundbreaking international collaboration with a memorandum of understanding. "It's more than an agreement not to shoot predators," Sergio explained to me on our last day. "It outlines our shared goals about conservation and will help us shape strategies to restore and protect the riparian and desert habitat of El Aribabi."

Jessica, who participated in the signing celebration at El Aribabi, says the event marks a new and hopeful outlook on conservation across international borders. "It is so inspiring to see people working together to make the world a more diverse and beautiful place," she says, "especially people divided by so many barriers: languages, lifestyles, and fences. Ranchers and biologists sitting together and talking about jaguars and ocelots, and excited about it. It's heartwarming."

At the tiny community of Mascarena, Sonora, two miles south of the border, the Santa Cruz flows northwest in a broad and shallow riverbed whose grassy banks darken under the spotted canopies of cottonwood, willow, and oak. Except for an increase in the number of trees, the scene has changed little in 120 years, based on an 1890 photograph taken near here.

I walk along this perennial stretch of river where time seems to have averted its gaze, and I think about how nature is, indeed, the untranslated word of God and how there are still holy men like Sergio and Carlos who speak for it. In *La Frontera*, Alan Weisman writes: "Every bend of the Santa Cruz is worthy of a gasp." Worthy of a gasp—yes. But I would add, "and a halleluiah."

Jessica Lamberton at El Aribabi

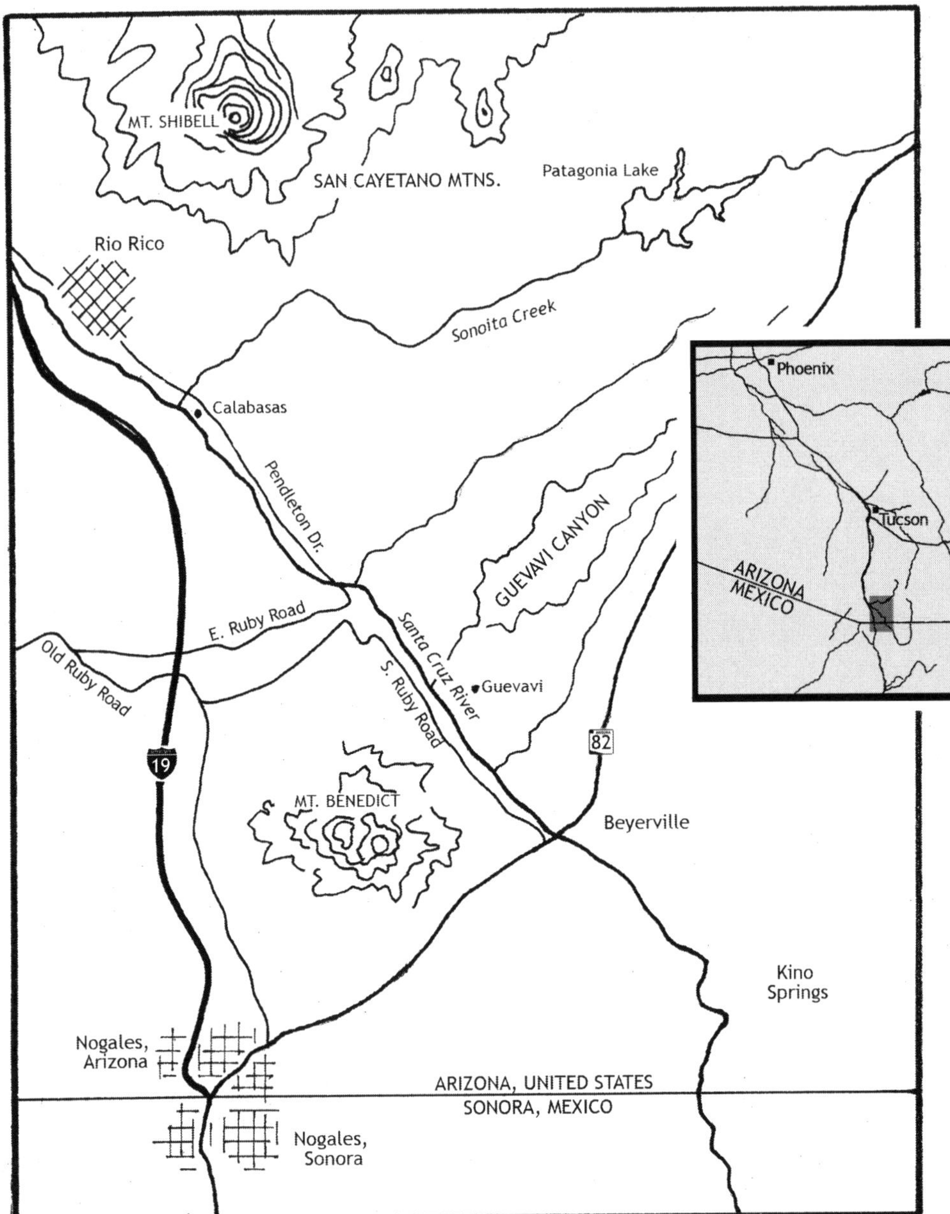

The Kino Springs Reach

4

Border Crossings

The Kino Springs Reach

Life is where change and
redemption are possible.
—Wendell Berry

Guevavi: Arizona's First Mission: October, 52 River Miles

An occasional car Dopplers along South River Road, which contours the Santa Cruz on our left. Melissa, my youngest daughter, who has joined me today, suggests that we duck and hide at the sound of tires on pavement. I know what she means; I feel it also. We're too exposed. Our tall frames are too easily seen, trespassing among the brittle, desiccated seep willow and cocklebur as if being alive were somehow offending the dead.

Mount Benedict wrestles for dominance of the western skyline, its buried pediments squeezing the aquifer enough to bring water above ground in the form of springs and surface flow, but not recently. Except for the occasional storm runoff, water hasn't flowed here since 1993, the beginning of our current drought. And I'm guessing it's been much longer for the springs. Where the river channel swings wide in its course along some low bluffs, laying up alluvial sediments in a field-size terrace, I imagine clustered pyramids of anemic cornstalks instead of the persistent seep willow and cocklebur. This could be the place.

To reach the bluff, Melissa and I climb a granite outcrop just beyond the terrace, gaining footholds in a rough seam that could once have been the runnel of a spring, and slipping under a barbed-wire fence marked "U.S. Boundary NPS." Melissa, noticing other signs on the fence, says, "There's nothing we can do here that would be legal." I, too, am bothered by the plethora of "Posted" signs on all the fences. I recall news items about gun-toting vigilante ranchers, and I've seen Border

Patrol and Department of Homeland Security vehicles, both marked and unmarked, cruising these back roads on the lookout for Mexicans on foot. But we keep going.

At the top of the outcrop, I know this is the place. At my feet, a single round socket in the rock articulates both a fist of stone and evidence of the river farmers. It's a three-hundred-year-old bedrock mortero, a grinding hole for corn.

I scan the high desert grassland spread out beneath the gunmetal Santa Rita Mountains in the north. Dwarf mesquite rise to eye level and stop, releasing my view—and then I see them, chocolate walls poking up from the thin, crooked trees, the jaundiced San Cayetano Mountains behind them. We walk through what had been Father Joseph Garrucho's entry into his courtyard (a clear path on the ground), turn right toward the ruins, and climb a slope of loose earth between two five-foot walls to enter the church's nave. There's not much here; the sun is still hot on my neck. These few adobe walls, bleeding back into the dirt from which they came, are all that remain of Arizona's first mission.

In 1701, Father Eusebio Kino established the Mission San Gabriel de Guevavi at the native village of Guevavi (from an O'odham word, *gi-vavhia,* which means "big spring"). Kino was keeping a promise he made to these people after his earlier visit to Tumacácori and Guevavi in January of 1691 when he first stepped into what is now Arizona.

The legendary archaeologist and historian Charles Di Peso tells the story of how a delegation of Indians met Father Kino and his hawk-nosed companion, Father Visitor Juan María Salvatierra, as the two traveled up the Río Altar on a tour of Pimería Alta to assess the loyalties of the native O'odham. Kino's and Salvatierra's superiors thought the O'odham might be in league with the Apaches and attacking Spanish settlements, something Kino seriously doubted. The delegation, a group of Indians from villages along the Santa Cruz River, including Tumacácori and Bac, sought out the Jesuit as their champion, carrying simple wooden crosses as a gesture of friendship. "Those crosses which they brought were tongues that spoke much and with great force," Kino would later write, "and that we could not neglect going where they summoned us with them."

Upon their arrival at Tumacácori, the priests found three ramadas—one for celebrating Mass, one for sleeping, and the third for preparing meals—which their hosts had raised out of mesquite branches for them. "These people made clear their intent to become allies of the Spanish," Di Peso writes. "It must have been a momentous decision for the natives, and a great deal of planning must have gone into it." He suggests the O'odham may have considered their survival as dependent on peacefully accepting the European intrusion, or that perhaps the natives recognized a material advantage in the way of new seeds, livestock, and farming technologies. I think their interest in an alliance with the Spanish may have been as simple as gaining protection from an ancient enemy of the O'odham: the Apache.

Either way, ten years after Father Kino first entered the Santa Cruz Valley at the request of the O'odham tribes, he made Guevavi the district headquarters, or *cabecera*, of his missions and Father Juan de San Martin became the mission's first resident priest. Father Martin immediately began constructing a small church and house but within a few months left Guevavi for "health reasons," which could have meant any sort of malady including malaria or Apaches.

For the next twenty years, missionaries visited Guevavi only occasionally. Then, in 1732, Father Johann Baptist Grazhoffer, a pockmarked and sandy-haired Austrian, reestablished Guevavi as cabecera, finished construction of the first church, and died the following year, probably from poisoning. Marriage customs of the Indians didn't mix well with Roman Catholic theology. At the time, O'odham marriage was without ceremony and fairly loosely structured. Husband and wife could separate on a whim and both were free to marry again. Promiscuity prevailed among both sexes, and polygamy was only a question of the husband's ability to support more than one wife. Historian Marshall Trimble puts it this way: "Apparently hard feelings developed between the German padre and his protégés. He resented their boozing and wenching, and they took exception to his piety. Somebody poisoned his dinner, and he died on May 26, 1733."

Perhaps the Indians better understood that sex is a language as much as prayer, that both are attempts at oneness, a communion with the sacred. Maybe the O'odham took exception to a man who denied himself half the words for speaking with God.

It wasn't until the mid-1700s under the productive oversight of the blue-eyed Sardinian, Father Joseph Garrucho, that Guevavi finally became the mission whose ruins Melissa and I are seeing today. It rose out of the baked ground to encompass an area about the size of a baseball diamond. The plans called for a rectangular church, fifteen by fifty feet, with three-foot-thick walls of sun-dried adobe, plastered with mud, whitewashed, and then painted with colorful decorations on the inside. A school, kitchen, and refectory among other rooms hugged the perimeter walls around an open courtyard. Standing high on this mesa, the church would have been an impressive sight for dozens of miles in all directions. And the visibility worked both ways. Off the southeast corner of the church's narthex, a circular tower, unusual for missions at this time, would have given sentries a perch to watch for approaching Apache raiders.

Melissa and I stand among three leftover adobe walls, the highest only about eight feet tall, all of them crumbling and chocked with rocks. A loose pile of horse droppings rests where an altar once held silver chalices. I remember hearing about archaeologists finding chicken bones and peach pits in the walls and how they were made by men and women's hands, some not so willing or careful. I think about how the mission must have appeared two hundred and fifty years ago. Its plastered walls shimmering in the heat rising off this corrugated landscape. Arizona's first White Dove of the Desert. Guevavi would have inspired more than native eyes.

The first time I visited Los Santos Ángeles de Guevavi was with a tour led by Tumacácori National Historical Park volunteer ranger Grant Hilden, who packed six of us in a low-clearance van and drove us swaying and bumping across this arroyo-cut ranchland. Grant showed us Father Garrucho's entryway and invited us into the church's nave. He told us that we were standing on two famous graves: Captain Juan Tomás Belderrain, who died at Guevavi after being wounded by Seri Indians, and María Rosa Bezerra Nieto, the mother of Captain Juan Bautista de Anza, the renowned Spanish explorer and soldier, and Belderrain's successor at the Tubac presidio. They're buried side by side. "Guevavi has ninety-eight burials that we know of," Grant said. "It was a hard place to live."

Father Garrucho found himself in the center of one horrific event that would end in more than a hundred deaths—natives, settlers, and two

Ruin of the Guevavi Mission

missionaries. On Sunday morning, November 21, 1751, as the sun rose over Pimería Alta, O'odham warriors began attacking and killing Spaniards in a revolt that swept from Sonora down the length of the Santa Cruz River valley to the northernmost mission near Tucson. The Indians ransacked San Xavier. At nearby Tubac, they stripped and burned the church. When a bloody Juan de Figueroa, the mission foreman at Tubac, stumbled into Guevavi to warn Father Garrucho, panic erupted. Garrucho's neophytes deserted him. O'odham not involved in the uprising (and some that were) fled into the mountains, fearing military reprisals. Father Francisco Xavier Pauer, the Jesuit assigned to San Xavier, escaped with the help of his O'odham neophytes, retreating with other refugees to Guevavi. But Guevavi wasn't safe either. By Wednesday of that same week, the padres with their livestock and church furnishings had abandoned the mission.

Spanish civil authorities blamed the Jesuits and their harsh treatment of the natives for the uprising. Father Garrucho was ordered to stand trial in Mexico City. One of those behind the rebellion, Pedro de la Cruz, alias "Chihuahua," a half O'odham, half Opata Indian, raised by a Spanish family, accused Father Garrucho of instigating the rebellion before officials summarily tried, convicted, and executed the rebel for treason. Father Garrucho kindly heard Pedro's last confession.

There's an interesting story that Grant Hilden told our tour group about the uprising of 1751, which may support Pedro's implication of the padre, although Grant admitted it's probably a myth. Pedro was the cousin of Luis Oacpicagigua, a shaman and captain general of the O'odham warriors who led the revolt against the Spanish. Luis had made Pedro his sergeant. "Father Garrucho was angry with Pedro," Grant explained, "because he had abandoned his wife and children so he could join Luis. Pedro carried a *bastón*, a stick that was a symbol of authority, given to important men in native communities. They were used in the same way that a signet ring might be used to prove you were someone of power. Supposedly, Garrucho was so angry he took Pedro's bastón and broke it over his head."

Father Garrucho was acquitted of any charges in Mexico City, but he never came back to Guevavi. Father Pauer, not yet ready to give up on the natives living along the Santa Cruz, took his place.

Melissa lays her hand on a mission wall that has stood in our desert for hundreds of years, and I think about how those same young hands have touched more ancient cathedrals in Spain. History collects in her pores. For seventeen short years she has lived upon this landscape of dirt and expectation, connecting with its immeasurable past, a tiny link in a vast human chain. Her mother and I conceived her during a hike to a high cleft of rock carved out by the wind. After her birth and before she could speak or crawl, we introduced her to the earth, placing her on its raw geology as if it were a religious rite, a celebration. She began to name the stones. We decided she would become a geologist, an anthropologist, a planetary historian, and so we raised her on the banks of this dry river named for the Holy Cross.

We may walk on this earth one layer at a time, but there are places where all the layers rise to the surface and we share the same elements of those long dead. And those yet to come. Places like Guevavi reinforce for me the truth that we are only a matter of time. "Why do we find it supremely pertinent," asks Annie Dillard in *For the Time Being*, "during any moment of any century on earth, which among us is topside? Why do we concern ourselves over which side of the membrane of soil our feet poke?"

I tell Melissa that we are standing over the bones of a man who was present as history swung on its hinges for Pimería Alta: Juan Tomás Belderrain. She stops in her tracks as if uncertain to take another step, not wanting to be disrespectful of the captain's resting place.

Captain Belderrain commanded what would be Arizona's first presidio, or Spanish garrison. As a consequence of the native uprising, the Spanish quickly established a strong military presence in the region, and by the end of 1752, Belderrain and fifty soldiers began building a stronghold at Tubac, about eighteen miles farther downstream on the Santa Cruz River.

With the presidio at Tubac, Guevavi flourished for some years after the uprising. The mission, however, would not survive long after a second storm—the Jesuit expulsion of 1767. With some notable exceptions, San Xavier's Father Francisco Garcés, the "Kino of the Franciscans," for one, it seems the Jesuits' brown-robed replacements weren't as dedicated to building thriving native communities.

On July 6, 1772, Father Antonio de los Reyes submitted his report on the condition of the mission, writing that "the village of Guevavi is situated on an open and fertile plain beside an arroyo with good land where the Indians cultivate their individual fields of wheat, Indian corn, other crops, and one small community farm. The church on the inside is adorned with two altars and a small side chapel with paintings in gilded frames. In the sacristy are three chalices, two dishes with cruets, one pyx, a ciborium, a censor, and a baptismal shell—all silver—and vestments of every kind and color and other ornaments for the altar and divine services." Father Reyes also mentions a native population of married couples, widowers, widows, and orphans that numbered eighty-six.

A year later, only nine families grew corn and beans and squashes along this bend in the Santa Cruz River. By 1776, their fields would lie fallow, never to be planted again.

Father Eusebio Francisco Kino died in 1711 on the ides of March without ever seeing the rise and subsequent decline of the first mission he started in Arizona. But his real legacy far surpasses one small mud-walled church on the rim of Christendom, a legacy that is at the same time both measurable and immeasurable. If there is a hinge between prehistory and history for Pimería Alta, Father Kino was its pin. Kino accomplished more than forty expeditions into the region, establishing twenty-seven missions, the future-glorious San Xavier del Bac among them. He drew the first accurate maps of Pimería Alta, the Sea of Cortez, and Baja California, and proved that the Baja peninsula was not an island as commonly thought. He discovered a land route from Sonora to California, what people would later call El Camino del Diablo, the Devil's Highway. His was the first recorded visit (in 1694) to the Hohokam ruins at Casa Grande.

Kino brought new farming and ranching technologies to the native people of the region, introducing them to wheat, cattle, horses, sheep, and goats. He also brought a new faith, introducing them to a new God. No one will ever make a full account of the O'odham lives he touched.

Father Kino "died as he had lived, with extreme humility and poverty," writes Father Luís Velarde, his friend of eight years and replacement at Dolores. Kino collapsed while celebrating Mass during the dedication services of a new chapel at Magdalena, where he was visiting from Dolores at the behest of Father Joseph Agustín de Campos, Magdalena's missionary. He died sometime near midnight amid the prayer vigils of the native people who loved him, and Father Campos buried him in the same new chapel they had just dedicated to San Francisco Xavier, Kino's patron saint. He was sixty-six years old. He had led a solitary existence in a foreign land, fraught with hardships despite his successes or failures. But this is the life Kino had asked for, and he never turned back from it. Perhaps he believed that God had redeemed for him fifty years he wasn't supposed to live. He responded by living it well.

On May 19, 1966, almost three centuries after Father Kino first arrived at Pimería Alta, researchers excavating the plaza in front of the Magdalena church uncovered a suspicious grave. The search for Kino's resting place had begun many years earlier, but each attempt had ended in failure. The historian and Kino biographer Herbert E. Bolton originally thought Kino's remains might have been relocated from Magdalena to the mission at Caborca, and in the late 1920s he dug around the church's foundations looking for them. Later, a Mexican anthropologist named Serapio Dávila opened a few trenches in front of the church of Santa María Magdalena but quit the excavation after finding an old cemetery full of unidentifiable human bones. In the thirties and forties, other researchers attempted to solve the puzzle without success. The most reliable record of Kino's burial came from the man who placed him in the grave, Father Campos: "he is buried in a coffin in this chapel of San Francisco Xavier on the Gospel side where fall the second and third choir seats." But where was his little chapel of San Francisco Xavier?

Then, in 1961, *Arizona Highways* devoted its March issue to Father Kino, celebrating the 250th anniversary of his death. The magazine stirred up even more interest in locating his burial place. Rumors, conflicting opinions, and false leads continued to frustrate researchers from both sides of the border. Finally, on June 30, 1965, Mexican president Gustavo Díaz Ordaz commissioned the right person for the job—Wigberto Jimenez Moreno, the famous ethnologist of pre-Columbian Mexico. Historians, archaeologists, and anthropologists from both Arizona and Sonora joined Moreno. Moreno's Kino team included Dr. Jorge Olvera, an art historian, and Professor Arturo Romano, an anthropologist. Both combed the historical archives. Father Cruz Acuña from Hermosillo searched archives of the Catholic diocese, while Reverend Kieran McCarty of San Xavier in Tucson sorted through the Franciscan records. Dr. William Wasley from the Arizona State Museum took charge of the archaeology.

After exploring miles of both microfilm and trench lines, Moreno's team narrowed down the probable location of Kino's grave to the plaza in front of the Magdalena church. Soon, workmen there exposed an adobe wall with the correct east-west orientation. When the wall turned a corner, the team suspected they'd found the foundation of a small building. Could this be the little chapel of San Francisco Xavier?

Father Charles W. Polzer, a Tucson Jesuit priest and ethnohistorian who researched and wrote extensively about Kino's life, said this about the moment of discovery: "Cautiously the crews followed the line of the adobe and boulder foundation. The sharp spades cut through the soil ever so carefully. A shovelful of earth spewed into the screen. Nothing. Then at the base of the trench fell a piece of skull, dislodged from the edge!" It was 4:45 in the afternoon.

Five days later, on May 24, 1966, Moreno's team announced to the world that it had uncovered the remains of Padre Eusebio Francisco Kino. The researchers left his bones on a dirt floor where they found them, and today a white domed monument in the center of a beautiful plaza adorns the site.

I've visited the padre's grave in Magdalena many times, admiring the humble man who cared so much for the native people in his charge. Even now, his bones convey his manner, lying in peaceful repose, his bronze crucifix, a symbol of his burden for men's souls, still marking the man as it stains his right collarbone green.

Calabasas: Mission of Gourds, 57 River Miles

Melissa and I walk downstream from Guevavi, following the right side of the channel and taking notice of raccoon tracks in the dried silt. *How long have those been there?* I wonder. Giant dead cottonwoods, some still standing but many lying like gray cadavers on the sand among the cockleburs, betray an inadequate subsurface flow. Deep roots mean nothing without deep water. The abundance of sapling cottonwoods encourages us, however. The river still desires a return to its former canopied magnificence, tentatively drilling wells with new trees.

Two weeks later, we're again walking the right side of the channel downstream from Guevavi. It's late afternoon and clouds bruise the sky, threatening rain. A gibbous moon, thin and stale as a communion wafer, rises in the east. Our intention is to hike from Guevavi to Calabasas, another eighteenth-century Spanish mission about five miles farther north, and then pick up the Anza Trail where it joins the Santa Cruz River near the Sonoita Creek confluence.

We walk the same path the stocky, dark-skinned, big-nosed Father Francisco Pauer (Spanish authorities were tedious about recording physical details) would have walked countless times as he ministered from Guevavi to the people in his charge along this part of the Santa Cruz River. Following his reassignment to Guevavi after the O'odham uprising, Father Pauer began enlarging the native population at a place he called "Calabazas" by gathering Indians from neighboring villages to this tiny settlement. On November 1, 1756, seventy-eight people arrived at Calabasas for baptism and marriage, and Father Pauer began calling the community a *visita*, or branch mission, to his *cabecera* at Guevavi.

I figure we can make Calabasas in less than two hours, despite the loose sand and frustrating lack of traction, despite the dry amaranth burs that cling to our socks and drill themselves into the tender flesh of our ankles—I wonder if Father Pauer ever tired of picking cockleburs from his black cassock! Darkness, however, could be a problem. This late in the year, the planet's shadow slips westward into sudden blackness rather than long hours of gloaming, and we will need the light to locate the mission's ruins.

At the Ruby Road overpass, the wind stirs. Ravens dangle and twist in the air and chortle. A single Say's phoebe calls mournfully from a naked willow branch, leaving vacant silences between each note. From the east, Mexican music rises and falls, undulating across the desert. Melissa and I walk without talking, our footsteps stirring an occasional grasshopper that fans its bright red and black wings and clatters to momentarily draw our attention before dropping back to the ground and vanishing. Clever animals, I think, offering a target with a flashy presentation that folds into camouflage at the last instant. The grasshoppers must be masters of near misses, an effective strategy to foil predators. Father Grazhoffer (Grasshopper?) also had a flashy way of attracting attention. Too bad he wasn't as skilled at blending into his environment as well. Perhaps the whole mission program could have learned something from these diminutive insects.

Where Melissa's steps uncloak a brace of grasshoppers and knock clods of sandy silt from a low bank, I notice how cattle hooves have broken whole sections of ground into puzzle fragments. Melissa could probably tell me all about the nature of the sediments, whether they're classified as silt or loam or silt loam or silty clay loam. She knows about dirt. She's wanted to be a geologist since before she could spell the word.

The Santa Cruz River has played a significant part in her life, living as we do on a small tributary arroyo of the dry river. In eighth grade, Melissa worked on a science fair project she called "Hydraulic Conductivity." The Central Arizona Project (CAP) canal had recently begun delivering Colorado River water to southern Arizona and engineers were devising ways to use some of the water to recharge our shrinking aquifers. Melissa's project involved measuring the rate at which water filtered through columns of various soil types, her intention being to locate the best places for this recharge. But what I mostly remember is driving her hundreds of miles all over the Santa Cruz Valley, digging holes wherever she pointed and collecting the dirt in buckets.

At the Southern Arizona Regional Science and Engineering Fair (SARSEF) that spring, Melissa won first place in her division, despite criticizing Tucson's water department for building a recharge station in the wrong place (the land was cheap there). The award allowed her a chance to participate in the Discovery Channel's annual Young Scientist Challenge in Washington, D.C. Four thousand applicants from all across the country including Hawaii would be narrowed down to forty finalists, based primarily on essays they would write about their projects.

In August of that same year, we learned that Melissa had made the first cut. She was among four hundred semifinalists. Then, on Sunday evening, September 9, 2001, a phone call came for Melissa. The four of us watched her without a word. "That's great," she said into the receiver, looking back at us across the dinner table with shining blue eyes. I raised my arms over my head. She and I were flying to Washington, D.C., in October.

The events of 9/11 nearly put an end to that year's competition, but parents, including Karen and me, encouraged the Discovery people to proceed with the science challenge. Although we were shaken by the attacks in New York and Washington, we all agreed the program was important to the children and that continuing with it was an appropriate response to the tragedy, a response of hope.

The Discovery people agreed.

A month later, forty middle-school children and their parents descended upon the plush Hotel Mayflower in downtown Washington, many of us arriving from small towns in states like Montana and New Mexico, New Jersey and Hawaii. Melissa and I were immediately impressed with the marble floors and gilded architecture, the Chianti and chocolate-dipped strawberries. We dropped our luggage in our room and took a walk, no destination in mind. When we looked up, we were facing the White House.

Over the following week, Melissa and the other science challenge kids competed in teams and as individuals to determine who America's top young scientists were. The parents, mostly to keep us occupied and away from our children, toured the area. So, while Melissa explored the back rooms of the Smithsonian Institution, examined plants collected by Lewis and Clark, and met people like Craig Venter, the force behind the Human Genome Project, I was bused with a horde of noisy forty-somethings to places like George Washington's plantation. I was so jealous.

By the end of the week, all forty kids had their names in the stars. Literally. Everyone had an asteroid named for them. Melissa's is "Minor Planet 15624 Lamberton." At the awards ceremony, one of the judges gave Melissa a discretionary prize because, he said, she has "a sharp eye connected to a sharp brain." It was the one award we had both already decided would be the best to win: two weeks in Hawaii to participate in TOPS (Toward Other Planetary Systems), an astronomy program on the Big Island associated with the Keck observatories on Mauna Kea.

I look at her now, and she smiles bravely. She's tied her jacket around her waist and her face shines with perspiration. She's tired but won't complain; she never does—whether she's joining me on my crazy endeavors or dealing with the frustrations and long hours of her own projects.

It was her two weeks in Hawaii with TOPS that really launched her interest in space science. While Karen and I and her sisters toured the olivine beaches, the volcanoes and sulfurous calderas, the giant koa forests of the Big Island, Melissa stayed up nights with astronomers and other students from all over the Pacific Islands, photographing and calculating the depth of lunar craters, locating Messier objects, measuring the spectra of stars. On the nights that I joined her above Hapuna Beach, the stars were palpable. A telescope seemed superfluous. The Milky Way, a band of stars breaching the curve of sky from Cygnus to Scorpio, could have been clouds of my own astonished breath.

After Hawaii and TOPS, Melissa's science fair projects, which previously had been grounded in soil studies and equations, left the planet altogether, although she continued to focus on dirt. In tenth grade, she created a mathematical formula to show a relationship between a soil's organic content and its climate in terms of moisture and temperature. (This time I dug holes all across southern Arizona.) "The equation establishes," she wrote in her conclusion, "that where specific climate conditions exist, terrestrial or extraterrestrial, life has potential. When we search for life, on Mars or elsewhere, this equation will tell us where to look."

Melissa's eleventh-grade project practically landed on Mars. This time she decided to duplicate Martian soil-climate conditions and see if an organism could grow in them. Relying on her TOPS connections to scientists at the University of Arizona, she gained access to a microbiology lab and the bacteria she needed for her experiment—a cold-loving "psychrophile" from the dry valleys of Antarctica, which arrived in a sealed glass ampoule as if it were a bio-toxin. Her soil, a "Martian analogue" chemically similar to soil found on the Red Planet, came from Hawaii. She chose five different moisture conditions each with five different temperature conditions, including a –80 degree Celsius deep freeze of cracking ice and frozen vapors, for the soil samples, two hundred test tubes in all, which she inoculated with the bacteria.

Once a week for eight weeks Melissa washed the bacteria from the red soil and measured their growth with a spectrophotometer. What she found was unexpected. She thought she should call NASA. The bacteria were thriving, but not in the kind of conditions where NASA would be looking for life on Mars. The bacteria liked it cold and dry rather than warm and wet. In fact, the colder and drier the better. NASA, she thought, should be sending spacecraft to the dusty margins of the Martian polar caps.

This year Melissa is working with researchers at the University of Arizona's Lunar and Planetary Lab on NASA's Phoenix project, a Mars polar lander "rising from the ashes" of the previous doomed Mars polar mission. Pat Woida, a UA scientist developing the imaging systems for the spacecraft, was so impressed with Melissa's science project that he asked her to join his team. She's trying to find ways to protect Mars from Earth microorganisms that may be riding along with the lander. I like to joke with her that she could send along some of her Antarctic bacteria and begin turning the Red Planet green.

From the sands of the Santa Cruz River to the soils of Mars, the planet of dry rivers. I told Melissa recently that she could be the first woman geologist on Mars if she wanted to. She corrected me, however, saying, "Dad, I could be the first *person* on Mars."

Melissa and I try to locate the ruins of Calabasas where they rest on the bluffs above the Santa Cruz River. I hold in my hand a copy of a drawing made by Charles Shuchard, a railroad survey artist, who sketched Calabasas as the mission appeared in 1854 when it was a hacienda and ranch. We want to find the artist's vantage point and compare the two scenes, 150 years apart. We eventually do—with some difficulty—climbing west of the river into a hillside housing development that faces northeast across a bend in the Santa Cruz. It looks like Mount Shibell is missing from the drawing, and then we realize, after having hiked back and forth across the valley, that we must line up the smaller peak with 6,007-foot San Cayetano Peak, the prominent feature of Shuchard's drawing, to make them one peak.

On the right flank of the peaks and below the distant blue-shadowed Santa Rita Mountains, new homes overshadow the ruins, whose blond

Ruin of Calabasas

adobe walls rise like termite mounds beneath a shining protective structure. Melissa spots cows moving near the ruins like the ones in Shuchard's picture. Some things never change. As I look out at the landscape, I can see the same line from the drawing that marks the course of Sonoita Creek as it flows west toward the Santa Cruz, although low mesquite trees have replaced the cottonwoods. The Santa Cruz Valley in the foreground, once a grassy pasture for sheep, is today matted with mesquite. The trees cover everything. Even the bluffs and the hills beyond them wear nappy sweaters of mesquite. Along the river itself, a few cottonwoods spread branches with moth-eaten canopies. The cottonwoods are spindly and anemic but persistent.

On February 8, 1857, a few years after Shuchard sketched this place, adventurer and explorer John Coleman Reid wrote in his journal, "If you will portray in your imagination a bottom covered with tall, golden colored grass, hedged by mountains whose sands glitter like metal, divided by a meandering stream a dozen yards wide and as many inches deep, this shaded by cotton-woods, willows, and musquites [*sic*], then a few hundred yards higher up another stream, a creek with less volume pouring in from the right, and in the fork an elevated rolling surface, you will have a view of Calabasas (Pumpkin, so called from an old adobe house, named from its color, which stands on the right bank of the river near the above noticed junction)."

Melissa and I cross the Santa Cruz and climb the bluff toward the same "old adobe house" of both Reid's pen and Shuchard's pencil. But unlike the ruins at Guevavi, we can't enter through the "gate" and step inside his church. Chain-link fence topped with a spiral of barbed wire bars our way, and we must satisfy ourselves by looking without touching.

We wrap our fingers around the rough galvanized metal and transfer some weight to the fence. Three dark openings—two doors and a window—stare back at us, the linteled rectangles evenly spaced along a long adobe wall maybe ten feet high. Even at this distance, we can easily see each adobe brick, still mortared into place after two and half centuries. In places, a lighter plaster smooths the mud-daubered exterior, hinting of the church's former grace. Here, we look at history and history looks back, but not with the disinterested eyes you'd expect from something that has seen it all. Something that has noted the passage of centuries as we might count the minutes, that uses the wobble of Polaris for a calendar. The place seems to say to us that the footprints we leave here matter as much as those that have already come to Calabasas, and those that are yet to come. It says, as the poet Richard Jackson writes, that we all drink from the same footprint of history.

We know that people lived here for more than three thousand years before the Spanish arrived. They farmed the fertile benches where two creeks, the Potrero and Sonoita, join into the Santa Cruz River. In January 1691, when Father Kino founded his missions at Guevavi and Tumacácori, he encountered communities of O'odham growing corn and squashes along this part of the river just as their ancestors had for millennia.

The Calabasas church Melissa and I see only as adobe ruins today probably began construction around the time King Charles III expelled the Jesuits in 1767 and the gray-robed Franciscans came to the Santa Cruz Valley. Apparently, the Jesuits started with some mud walls and the Franciscans finished with a straw and earthen roof.

In its glory, the church would have been a reassuring sight. Like the mission at Guevavi, its high white walls would have stood above the surrounding landscape, creating a comforting place to rest your eyes as you worked the nearby fields. Inside, the church's bright plastered interior with its paintings of saints, its choir of young native voices with flute and guitar accompaniment, would have provided a place of refuge, a modicum of peace and comfort at a time of uncertainty and death. The church would have embodied the Christian heaven brought to earth.

But even these powerful symbols of a new faith couldn't thwart the Apaches. As the raiders once again began stepping up their attacks and taking a greater toll on the small villages, people continued to gather together for protection. Guevavi was abandoned. Losing the Spanish military garrison at Tubac, which moved to the new presidio at Tucson, was another blow. Settlers, petitioning for the return of the presidio to Tubac, wrote on November 24, 1777, "Since the fort was moved to Tucson, these towns and missions have experienced some casualties; so much so, that they have been obliged to burn the town of Calabasas." Over the next fifteen years, most of the people at Calabasas would flee to Tumacácori, and the Franciscans would convert the church into a stock ranch for the Tumacácori mission. But even the ranch didn't survive to the end of the century.

In 1806, Juan Legarra, the forty-year-old Indian governor of Tumacácori who had been born in Guevavi, asked the government of Sonora for the land surrounding Calabasas for his people. He mentioned Calabasas specifically for use as a ranch. The Spanish granted him title in 1807. The history of the place is sketchy at this time, but it seems that Calabasas, under Juan Legarra, continued at least intermittently as a ranch until the 1840s. Most of the Franciscans were gone, the Spanish missionaries having lost favor with the new government after Mexico gained its independence from Spain in 1821. (On April 16, 1834, the Mexican government decreed an end to the mission system, although Sonora gained a reprieve and a few Franciscans remained in the region.) Sonora had slid into political strife first with skirmishes between would-be governors Manuel María Gándara and Jose Urrea and then later between Gándara and Ignacio Pesqueira. And the Apaches, taking advantage of the Spanish withdrawal and subsequent civil chaos, redoubled their raids, killing settlers and forcing thousands more to abandon their pueblos, mining camps, and ranches.

Two years after the Mexican dictator Santa Ana decreed in 1842 that the mission lands be sold, Gándara's brother-in-law, Don Francisco Alejandro Aguilar, paid $500 at auction for the grazing land surrounding Tumacácori, Calabasas, and Guevavi, but neither Gándara nor his brother-in-law did much with the Calabasas ranch by the time John Bartlett visited there in 1852. The boundary commissioner described Calabasas as "the ruins of a large rancho."

Finally, between 1853 and 1854, Gándara's business associates, two Germans named Champon and Hulsemann, converted the old church into a ranch house with an enclosed yard, large corrals, and a flour mill and constructed a new barracks and mess hall to the north of the "hacienda." The Germans had employed a number of Mexican workers and their families, stocking the ranch with cattle and goats but mostly with sheep. This is the scene the artist Charles Shuchard drew when he arrived at Calabasas with the railroad survey team of Andrew Gray in April 1854.

Melissa says the mission/ranch house is smaller than she thought it would be. I think, at twenty feet by sixty, the place has twice the area of the six-hundred-square-foot home she's lived in for seventeen years. Since her birth, since the first days she could crawl over the ground and find her "precious rocks" to press into the hands of her mother for safekeeping, Melissa has called our converted garage and few acres of desert west of Tucson her home. The desert plot is the same land on which her mother was raised. It's all Melissa has known and, until recently, she hasn't wanted to live anywhere else. (Now, with college on her horizon, she thinks being near a California beach would be nice.) Even living with two older sisters, the three of them triple-bunked in one tiny room, has triggered only a need to be different, which she expresses most profoundly with her black belt in karate. If living all these years in one small place has affected her in any way, it certainly hasn't limited her.

There are other, more costly, responses to living in one place. One of them is the need for acquisition, filling the vacancies in our lives with substance as if we could add to a meager existence by hoarding mammon like some dead Egyptian pharaoh buried in a tomb.

The writer Kathleen Norris, who knows something about a cloistered life, says that a person can be forced inward by the spareness of what is outward. This is a common thread in the stories of life along the Santa Cruz River. Some people react to lean circumstances by reaching and clutching; others respond by looking inside and, perhaps, finding clarity. For Melissa, this clarity means learning to become native to her place, fitting herself to the landscape rather than conforming the landscape to her wishes. Like others who came before her, she's discovering how to live within the limits of this desert—knowledge she can pass on to her own children.

When Major Enoch Steen arrived at Calabasas on November 27, 1856, with his four companies of United States Dragoons and a complement of blacksmiths, carpenters, wagon masters, cooks, and laundresses, Manuel Gándara's ranching enterprise had already collapsed. Steen liked the place for its abundant water and grass, its proximity to the Sonoran border for supplies, and its strategic location for protecting settlements from Apaches. He called it, unofficially, Camp Moore.

The U.S. Army's encampment at Calabasas had a seemingly overnight impact on the region. Now, believing that they were safe from the Apaches, settlers returned to the abandoned ranches and farms, claiming lands once held by Gándara and others. Of these settlers—often referred to as "squatters"—some were among the territory's most illustrious pioneers. First, there was William Mercer, deputy collector of New Mexico, who lived at the adobe hacienda where he ran a supply store and kept his revenue office. In late 1859, the hacienda had another resident, a widower named Elias Green Pennington.

Pennington had come to Arizona from Texas with his twelve children after the death of his wife, attempting to make a life in the Santa Cruz Valley by farming its rich bottomlands. Unfortunately, a new life in this land also meant becoming entangled with the landlords, the Apaches. One famous story tells how, in early 1860, his twenty-two-year-old daughter, Larcena, barely survived being captured, speared, and beaten by Apaches after she arrived at her husband's lumber camp in Madera Canyon. Only a year later, Apaches ambushed and killed her husband, John Page. Elias himself and two of his sons suffered the same fate in 1869. Today, a short, narrow street (with a wonderful little coffee shop) in downtown Tucson carries the Pennington name.

Part of the problem for the settlers at Calabasas was that shortly after setting up Camp Moore, Major Steen moved the whole outfit to a second post twenty-five miles up Sonoita Creek. But even the new Fort Buchanan wouldn't remain through the beginning of the Civil War. Apache warriors, without the U.S. Army to keep them in check, once again invaded the Santa Cruz Valley, attacking settlements with impunity.

Another illustrious pioneer squatter, Pete Kitchen, had built his first ranch at nearby La Canoa only to abandon it after the troop withdrawal. When the Civil War ended, however, he found the area around Calabasas more to his taste and moved there to construct "El Potrero," his five-room, ranch-house stronghold, near Potrero Creek and the Santa Cruz River. Kitchen's fortified hilltop home, with its two-foot-thick adobe walls, high ceiling with roof access to parapets and portholes, held off the Apaches for many years. An 1872 *Arizona Citizen* article reported that Kitchen's farm was located in one of the most exposed places for Apache raids, but "instead of being frightened or discouraged by these bold and numerous attacks he seems only more determined to stand his ground and take his chances. The Indians have learned to their sorrow that in him they have no insignificant foe."

In 1878, just before his death at age seventy-seven, Manuel Gándara sold his Calabasas hacienda to a New York entrepreneur out of San Francisco named Charley Sykes. Sykes, in turn, sold the property to an eastern firm, which named the outfit the Calabazas Land and Mining Company and hired Sykes as its director. Always the salesman, the self-proclaimed "Colonel" Sykes produced several enthusiastic brochures about Calabasas, some with illustrations showing paddlewheel steamboats plying the Santa Cruz River!

Sykes did build an elegant brick hotel with verandas, black walnut furniture, and Brussels carpet, and he hired girls out of Boston as hostesses to serve elaborate meals of mutton roasts, lobster, and salmon. The two-story Santa Rita Hotel, called the finest hotel between San Francisco and Denver, stood above new stores, boarding houses, gambling casinos, saloons, dance halls, and an opium den—all pushing northward along the banks of Sonoita Creek. Notwithstanding its finery, "Pumpkinville," as the *Tombstone Epitaph* called the town of 150 people, had become the notorious "Hell's Hollow."

This was the colorful place James Cabell Brown wrote about in his funny (and admittedly fictitious) 1892 book, *Calabasas; Or, Amusing*

Recollections of an Arizona City. Upon his arrival there, when the Santa Rita Hotel was only a foundation, Brown compares a drawing of the "flourishing city" with what meets his eyes: "It was a good place to take a Turkish bath," he says, "for the sun beat down with an intensity unequaled elsewhere; but to perspire in Calabasas was a shocking waste of whiskey. . . . It might have been utilized as a burial place for the millionaire dead. In its climate and alkaline soil the bodies would have been incorruptible, and an incorruptible millionaire, dead or alive, would be worth a pilgrimage to see."

With the New Mexico and Arizona Railroad coming down Sonoita Creek on its way to its Nogales connection with Guaymas, Charley Sykes and other promoters expected Calabasas to become the port of entry to Mexico and a thriving gateway metropolis surpassing Tucson. Unfortunately, railroad officials spurned the inflated land prices and moved their customhouse and railroad siding ten miles farther south to the border town of Nogales. Founded in 1882 with the arrival of the railroad, a town named for walnuts had supplanted the one named for squash.

In the end, the Santa Rita Hotel, a symbol of Charley Sykes's hopes and his vanity, would become a rancher's hay barn before burning down altogether with the rest of the town in 1927. Yet, from these ashes, originally plowed under to enrich a cotton field, would arise Rio Rico, today's "Rich River" community of two thousand people. It seems the "Town of Gourds" has a much better claim to the name "Phoenix."

Evening arrives as Melissa and I depart Calabasas and follow an old two-track though a dense mesquite bosque along the east side of the Santa Cruz. We should find the Anza Trail somewhere between the mouths of Potrero and Sonoita creeks, but I'm not sure exactly where. Karen and our middle daughter, Kasondra, who are waiting to meet us at the trailhead, will be getting worried.

We leave the trees behind and head off west across a blasted Santa Cruz floodplain. Dead snags look sketched and skeletal; the ground buckles beneath our wooden legs. My nose tells me that we're downstream from the Nogales wastewater treatment ponds. A low dark line of trees in front of us indicates where the facility's discharge flows into the Santa Cruz River. I cut an angle more northwest toward a railroad trestle, and it's there that we find a confluence of effluent and sand.

The Santa Cruz River, now given a swift and turgid life by the citizens of Nogales on both sides of the border, drains Potrero Creek, which, a few miles south of here, drains Nogales Wash—this being the route Juan Bautista de Anza and his colonizing expedition traveled after crossing into present-day Arizona in 1775. Somewhere near here is a marked portion of the new historic trail. Melissa suggests that we cross the water by climbing across the trestle, but I'm not ready to make that commitment yet, because I still want to find the mouth of Sonoita Creek. Soon, however, I begin to regret not taking her advice.

Lower Sonoita Creek, 63 River Miles

Last August, under the angry blister of the sun, I walked in the sodium-white light following Sonoita Creek to where it empties only sand into an effluent-charged Santa Cruz River. Red cattle watched me from stunted mesquite along the banks. Thin, drought-pruned cottonwoods offered no canopy, no shade. These were not the large groves of "cotton-wood trees of gigantic size" that boundary commissioner John Russell Bartlett wrote about after arriving here in 1852. Nor were they the cottonwoods Charles Shuchard drew in 1853. And yet I wasn't surprised when a gray hawk suddenly appeared, circling overhead with pewter breast feathers burnished against a pale sky. The bird perched in one of the wasted trees to note my passing with dark eyes.

Recently, I had hiked most of the length of Sonoita Creek below Patagonia Lake with Michael Bissontz, bird-watcher extraordinaire, who has been monitoring birds like gray hawks here for ten years. He wanted to check on the status of a nesting black hawk, another important riparian raptor, and asked if I would like to join him. I first met Michael in Mexico at Centro Intercultural de Estudios Desiertos Y Océanos (CEDO), an intercultural education and research station near Puerto Peñasco, Sonora, where the Tucson Audubon Society was holding their annual Family Tidepools Institute. My wife and daughters called him the "bird man" and were impressed by his intensity and focus. "He must be able to see birds differently than most of us," Kasondra said once when I asked her what she thought of him.

Michael not only sees birds differently than most people, he sees more of them. He stopped counting after he hit 500 a few years ago but guesses he's close to 850 species. (My list shows a measly 246.) He's traveled as far as the Arctic plains of Barrow, Alaska, and he regularly ventures deep into Mexico to add more species. I spent a week with him in the Sierra Madre of Chihuahua, searching the back roads of Madera and Río Papagochi and the trails of Basaseachi, Mexico's second highest waterfall, for rare Mexican specialties like the Aztec thrush and the beautifully ornate eared quetzal. We found both of them and more than seventy other birds.

Michael is enraptured with birds; they are his greatest passion, a passion that took hold of him after he began working at a camp for emotionally troubled youths near his home in Buffalo, New York. Borrowing a pair of old opera glasses from his father, he learned to identify the birds around the camp and to teach what he learned to the children. Working on an assembly line at a General Motors plant became a dead end for him, so he went back to school and studied education. After graduation, he moved to Zuni, New Mexico, to teach Navajo children on the reservation. "I took the first available job to get out of Buffalo," he says. Soon, he discovered Tucson's bird-watching community, which drew him repeatedly into southern Arizona. Today, he lives in Tucson, still teaching children with emotional problems, and spends all of his free time walking the region's riparian corridors in search of feathered bodies. It's as if the birds called him here.

Michael and I drove across the Santa Cruz River at Rio Rico at seven in the morning. At a pond on the right side of the road, black ibises, beaks down, probed the shallow water like a scene lifted from the walls of an Egyptian tomb. Michael turned his truck east into a corduroy of mesquite hills and blooming shin dagger, paralleling a road just south of Sonoita Creek. Holding his Bausch and Lomb Elite binoculars in both hands and wearing a worn ball cap over his thinning red hair, he stepped carefully along the descending trail toward the creek. I followed. He walked a bit awkwardly, goose-stepping his long legs over the loose, rocky ground in a way I've noticed with other bird-watchers, who keep their eyes on the trees and sky rather than on the terrain in front of them. Before I could smell water, Michael began listing birds—whether he'd seen them or heard them I wasn't always sure: blue grosbeak, vermilion flycatcher, canyon wren, Lucy's warbler, yellow-breasted chat . . . I scribbled them into my journal as he spoke their names.

The chats were living up to their namesake, flying around and chattering wildly in the trees. And they continued to do so as we walked upstream, crossing an occasional slip of running water where the stream surfaced through the sand. The flowing water surprised me—and not only because it was early June. The water surprised me because most of the West had been suffering an extended drought of half a dozen years, and when water did come to the desert it was rarely a green ribbon but a churning chocolate mass in the afterbirth of a storm. Something that didn't stay for long and took huge sections of the landscape with it.

Sonoita Creek wound through close canopies of cottonwood and galleries of willow and hackberry, their shade so deep it felt like a wet bandanna on my face. I imagined that this was the way the Santa Cruz River near Tucson once smelled and felt in June: humid but pleasant, a shade-green place to rest your eyes, chipping with birds and clicking with insects, instead of the 105-degree, zinc oxide cast that currently weighs against your skin while hiking the cement-walled channel.

Where the creek slowed and pooled, sifting through a fencerow of rushes, a leopard frog breached a mass of bright watercress and shoals of topminnows rippled the water's surface. Topminnows and leopard frogs—two animals once very common in Arizona. The Gila topminnow, a tiny olive-bodied fish, was first collected and described in 1851 from the Santa Cruz River near Tucson. Fish in the Santa Cruz? It sounds like a joke: What? Sonoran rockfish? Sand trout?

Some yellow-finned males, black in courtship, darted through the shallows to defend territories from other, "lesser" males. These pale males, however, weren't completely left out of the action. Courtship and mating can be distracting as the peak of the breeding season approaches and as many as 98 percent of females become pregnant. The subordinate males, patient as ever, wait for opportunities to slip in and mate with unguarded females. The sly males even have an advantage for this kind of strategy—a longer sex organ. Who says size doesn't matter?

A mile or so upstream, a concrete buttress rose out of the sand and stood alone, a crumbling and yellowed monolith like the keystone of some lost arch. I recognized the buttress from similar ones I'd seen on the San Pedro River: the remains of an old railroad bridge, part of a railroad line shaped like a giant question mark that in 1882 first connected Tucson and Benson with Nogales and Guaymas, Sonora. Years ago, I hiked along its deserted grade near Fairbank to see my first gray hawks, and now, just beyond the buttress, I heard their descending alarm calls.

Historically, gray hawk populations probably centered on the Santa Cruz River. In 1937 ornithologist Arthur Cleveland Bent found them "quite common" near Tucson, and until the 1960s they nested along Tucson's Rillito River and Tanque Verde Creek. Today, the hawks have retreated to a few places like Sonoita Creek where they migrate each spring to breed and raise their young.

While the gray hawks screamed at us, Michael pointed to a large sycamore and a nest high in its branches. On the nest sat a large dark bird—a rare black hawk—that immediately dropped from the branches, unfurled its wings, and melded into the shadows of the surrounding trees.

Sonoita Creek is a keystone to the future of the Santa Cruz River. Hope for the Santa Cruz lies here. And not only with Sonoita Creek but with other vestige desert streams as well. Davidson Canyon, Cienega Creek, and even Sabino Creek, all part of the Santa Cruz watershed that courses through the Tucson Basin, retain native fishes with complements of other key species like leopard frogs and Mexican garter snakes. Other important wet relicts are Arivaca Cienega and Honeybee Canyon, but even some dry washes, the so-called xeroriparian habitats, have much to offer. These arroyos aren't dead rivers; they are hopeful rivers. The dry West Branch of the Santa Cruz River still maintains the highest biodiversity of the Santa Cruz River floodplain near Tucson and includes two species not known to live anywhere else in the Tucson Basin. Along the wash's sun-baked corridor, giant spotted whiptail lizards race their frenetic shadows through dead-blond grasses and narrow-mouthed toads rest inside rodent burrows like forgotten Buddhas.

Charles Shuchard drew three pictures 150 years ago of this part of the Santa Cruz River. Although his scene of the Sonoita Creek–Santa Cruz confluence has changed somewhat, his depictions of Tumacácori and Tubac look as though he might have drawn them yesterday. I admit that I'm an optimist, a hopeful romantic even, but I imagine that if the Santa Cruz ever recovers some of its former vitality farther downstream near Tucson, it will be the tiny, seemingly insignificant, vestige "species banks" like Sonoita Creek that will transfuse new blood into the river's arteries.

Downstream from the mouth of Sonoita Creek, Melissa and I search for a place to cross the river. Neither one of us wants to remove our boots and wet our feet. The sun has left a bloody hemorrhage low in the western sky, and shadows deepen to the color of gravestone. The water smells rich and potent. Here, the river's dark vein nourishes everything, from the ribbons of watercress that cling to its margins to the great cottonwood galleries of Tumacácori and Tubac, before it rushes toward its sandy extinction.

A great blue heron, what Barry Lopez calls "that river monk," shakes out its gray cassock of feathers and lifts itself out of the channel on monastic wings. We've been hiking for three hours since Karen and Kasondra dropped us off at Guevavi, nearly twice as long as expected. I'm sure the Anza Trail is on the other side of the river; this side has become a slog through gravel swales, boulder piles, and tangles of seep willow and uprooted trees. We finally find a possible crossing where a thin cottonwood has fallen across the river. Melissa makes it easily, walking the log like a gymnast on a balance beam, slender arms lifted, blonde hair flying. I throw her all my gear: knapsack, notebook, wallet, hat, and sunglasses. My right foot plunges in with my first step, and Melissa says, "Careful. I'm *not* rescuing you if you take a swim."

Melissa's youth shouts in defiance of her mortality, as it is with all newcomers who step above the membrane of soil. I, however, am closer to the other side, to joining the multitude of witnesses who wait beneath our feet. Here, they remind me of it. In my bones, aching from exertion. In my throbbing vessels. In my thin and sun-creased skin. In this place of communion I walk to the edge of myself.

When I spot the Anza Trail marker a few dozen yards away from the river, I raise my arms over my head. Success! Now, even in the dark, we can make it to the trailhead.

As Juan Bautista de Anza's expedition of San Francisco Bay colonists passed this way on October 15, 1775, Father Pedro Font, the Franciscan monk who was traveling with them, left the group with an escort to say Mass at Calabasas. "Because this place is a region made dangerous by enemies," he wrote in his diary, "the commander did not permit Mass to be said in camp." By now, the mission at Guevavi was in ruins. Spanish authorities had arrested, deported, and jailed the Jesuits. Father Francisco Pauer was dead, having slipped away on his forty-ninth birthday as a prisoner in Cádiz. Revolts and Apache raids and European diseases were filling cemeteries to the extent that the natives probably associated the presence of a priest not with a new way of life but with the sacrament of last rites. Since Father Kino's shadow first fell across this river, the entire Santa Cruz Valley has reeled and convulsed with the dying.

But this wouldn't doom the river farmers. They wouldn't be a people relegated to ruins, abandonment, and scholarly theories. Death throes share the same bed as birth pangs. Death and birth occupy the same sacred space. Both are passages. Both have a painful, messy commonality, and it's not always clear what the outcome will be.

The future is always dark, they say, as dark as the grave—and the womb.

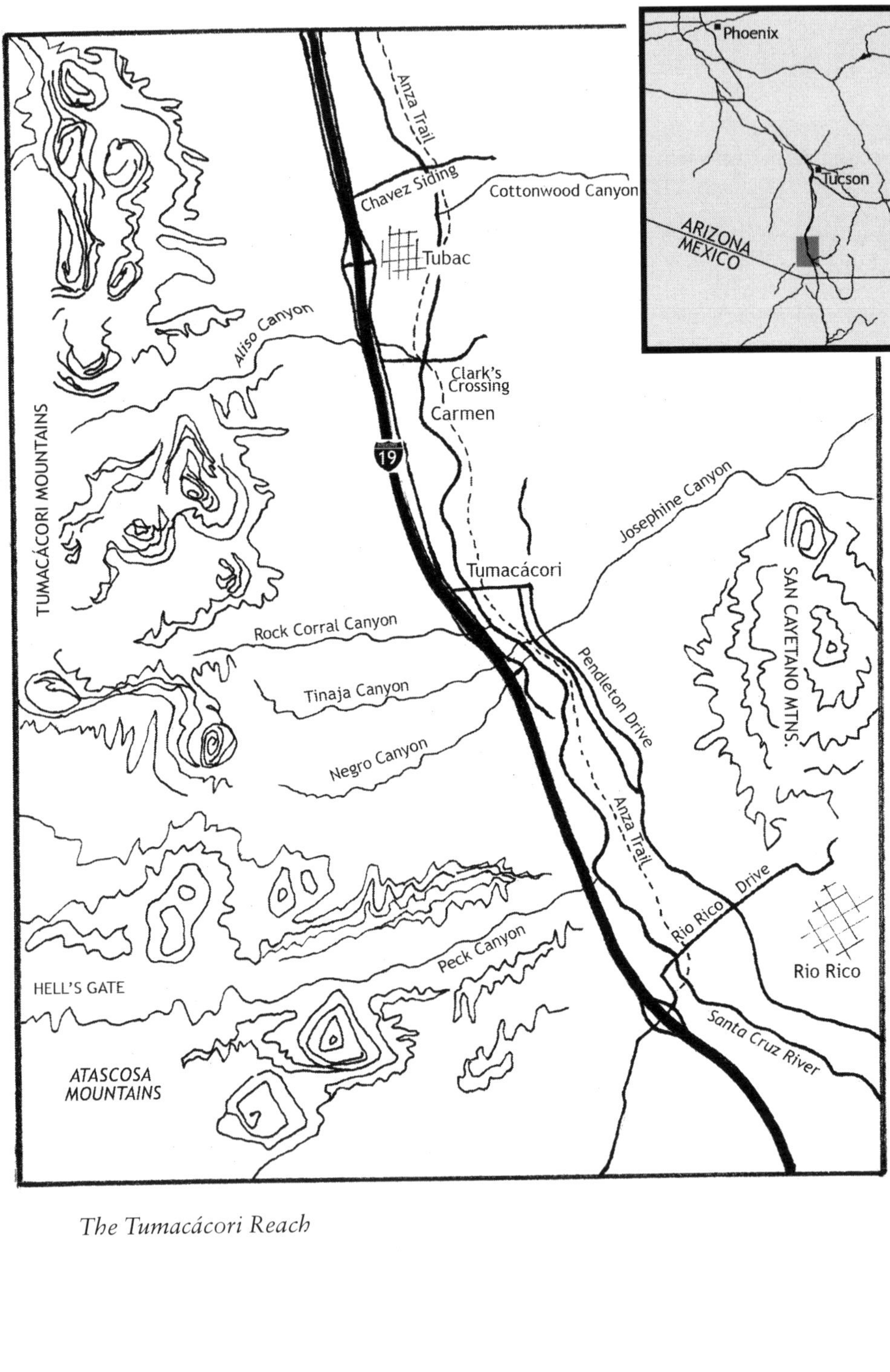

The Tumacácori Reach

5

A River That Is

The Tumacácori Reach

Rio Rico and the Anza Trail: May, 64 River Miles

Nearly seven months after Melissa and I hiked from Guevavi to Calabasas and joined the Anza Trail at Rio Rico, we return again. It's the end of May and school has let out for the summer. Karen, Kasondra, and Jessica have come along. Also with us is my friend Richard Shelton, University of Arizona Regents professor, writer, and poet whom I first met more than sixteen years ago when I walked into his creative writing workshop. Since then, Dick has been my writing mentor and hiking compadre; he and his wife, Lois, have grown as close to Karen and my daughters as family. My daughters, in fact, consider them godparents.

Dick wears blue jeans and a denim long-sleeved shirt. A white cowboy hat covers his tiara of shining hair. "I tried on my cowboy boots," he told me earlier, "but it seems my feet have swelled up over the years." Dick will be seventy-two next month.

We're all wearing boots and hats and jeans, as suggested by Marge Izzo who waves to us as Dick pulls his Dodge Caravan into the trailhead parking area. I've arranged to meet Marge and her husband, Ron, here where the newest section of the Anza Trail joins the Santa Cruz River on its way north through Tumacácori to Tubac about thirteen miles downriver. Ron is bringing the horses.

I contacted Marge and Ron Izzo of Arizona Horseback Experience a couple months ago after I decided that on horseback would be a great way to see this part of the Santa Cruz River, where, during the birth pangs of our country, Juan Bautista de Anza led his group of colonists on their way to San Francisco. Karen and the girls were all for a full day of horseback riding, and Dick agreed to come along too.

"They're on their way," says Marge, a bubbly redhead and principal organizer of our expedition who met Ron twenty years ago on a trail ride and has been at his side ever since. After Ron and his partner, "Wild Bill" McClain, arrive and unload the horses, we mount up under swirling barn swallows and the alarm screams of gray hawks. I see the birds—both species of which are significant for me—as a good omen. Marge takes the horse trailer to our takeout at Tubac and, with Ron leading the way and Bill taking up the rear, the six of us allow our mounts to jockey for position, working out their personality quirks with each other in terms of who will be nosing the rump of whom. The horses carrying Karen and the girls fall in line behind me, riding on Joshua. Dick's horse, a dark gelding named Willy, insists on taking the lead.

The trail, which Melissa and I had previously hiked from Guevavi and Calabasas to this point, now continues north, its sandy track marked periodically by signs that read "Anza Trail." In 1990, Congress authorized the Juan Bautista de Anza National Historic Trail, establishing the route Anza created in 1775–76 to open an overland passage for colonists and supplies to boost European settlements in Alta California. The twelve-hundred-mile trail stretches from Nogales to San Francisco and is the first of its kind in the West. And because Anza began his expedition in Mexico, plans are being developed to connect the remaining six hundred miles from Culiacán, Sinaloa, to Nogales and create the first international historic trail in the world.

Crossing from Mexico into what is now Arizona, Juan Bautista de Anza with his Franciscan companion Father Pedro Font and their party of colonists made their thirteenth camp at Las Laguanas, a cienaga just north of the present Saint Andrews Episcopal Church in Nogales. From there, they headed north following the Nogales Wash and Potrero Creek toward Calabasas, where the group picked up the Santa Cruz River coming in from the southeast. At this point, they began tracing the Santa Cruz, passing the village and church at Tumacácori, which would contribute a small herd of cattle to the expedition, and camping at Tubac. Continuing on to the mission at San Xavier del Bac, whose magnificent church was still twenty years from completion, and then Tucson, the colonists finally left the river near Picacho Peak. The group camped about five miles northwest of the Casa Grande ruins where Anza and Father Font visited to compare the descriptions and measurements first noted by Father Kino eighty years earlier. (It was here that Font wrote down the "Bitter Man" legend of the Pima Indians, the first record of the creation of Casa Grande.) Over the next seventeen days, Anza led the expedition westward along the Gila River, stopping to camp at several friendly Pima and Yuma villages, before crossing the Colorado River near present-day Yuma and continuing through southern California. After a few months' rest at Monterey, the colonists would complete their eighteen-hundred-mile journey at San Francisco nearly a year after leaving Culiacán. Along the way, they recorded nine births and four deaths—but only one death after leaving Tubac: a woman who died in childbirth.

Juan Bautista de Anza left the expedition at Monterey after first traveling to the San Francisco Bay area to choose sites for a presidio and mission. Because of his success, upon his return to Mexico City he would become commander of the military in Sonora and then governor of New Mexico, living in Santa Fe from 1778 to 1787. In the fall of 1788, Anza became commander of the presidio at Tucson, but it would be a short assignment. After reviewing the soldiers at Tucson, this remarkable man, whose generous nature affected the lives of so many, returned to his home at Arizpe and died unexpectedly on December 19, 1788. He was fifty-two years old.

His father, Juan Bautista de Anza Sr., had come from the Basque country of northern Spain to New Spain in 1712 at the age of nineteen where he explored an interest in mining in Chihuahua and Sonora. In the early 1720s, he joined the cavalry and quickly rose through the ranks, coming to the notice of Antonio Bezerra Nieto, captain of the presidio at Janos, Chihuahua. Apparently, someone also came to the notice of the young Juan Bautista. Not only did he make lieutenant at Janos, but he married the captain's beautiful daughter, María Rosa. In 1726, he became captain of the Spanish presidio at Fronteras, Sonora, where María gave birth to Juan Bautista, junior, in 1736. Unfortunately, the child would never really know his father. A few years later, Apaches killed him a short distance from the mission at Guevavi, where the grave of his wife María Rosa lies today.

Still, Juan Bautista the son followed closely in his father's footsteps. At sixteen, he joined the Spanish militia, earning a cadet position in the presidial cavalry three years later. At twenty he made lieutenant, proving his prowess as a frontier soldier and commander against the Apache and Seri Indians. At twenty-four, after Seri Indians killed Juan Tomás Belderrain, Juan Baustista became the second captain of the presidio at Tubac, holding this post for more than fifteen years until after his colonizing expedition to San Francisco.

This overland route from Sonora to Alta California, in fact, was originally his father's idea. In 1737, the elder Juan Bautista had requested permission from the viceroy of New Spain to search for the route, but he was killed before undertaking the mission. Spain had for years wanted to secure the Pacific coast against Russian and British incursions, but its small ships couldn't master the difficult sea passage with enough cattle and colonists. By the 1770s, Spain had only a few small settlements in Alta California, amounting to two presidios and five missions between San Diego and Monterey with a total population of about seventy Europeans. And to the north of Monterey, the Spanish had learned, was another much larger port, fed by a great estuary of the newly named Río San Francisco. Spain needed a land route to California if it wanted to ensure its possession.

Captain Anza personally financed the first exploratory journey in 1774 to prove that a route existed, charting along the way places for water and pasture, and befriending native tribes who might assist a larger expedition. With thirty-four men and dozens of wagons of equipment, cattle and horses, Anza left Tubac in January and traveled to the new presidio at Monterey, California, returning to Tubac by May. Now, with the viceroy's commission and a new rank of lieutenant colonel, Anza immediately began organizing a second larger expedition in Mexico City to colonize the San Francisco Bay area.

We struggle with the first river crossing on the Anza Trail, the horses balking at stepping into the deep-running effluent. The horses sniff the water and blow, backing away from the bank despite the encouragement of our boot heels and words. Apparently, our horses don't understand that all water in the desert is holy. Eventually, the more skittish horses follow the braver ones and after twenty minutes we gain the opposite side. I'm hoping we don't have too many such crossings.

The trail takes us through open mesquite and high, dead grasses where the midmorning sun presses down on us, releasing the smell of oiled leather and horse sweat. Gnats begin to orbit my ears and tickle my nose and lips. But as we enter the dense trees of the mesquite bosque, the temperature suddenly drops ten or fifteen degrees. It's like passing through a doorway into a darkened room with swamp cooling. Huge

Mexican elder join the mesquite, the former hung with clustered bouquets of creamy flowers.

Back in the open again, we take our first break, dismounting and allowing the horses to graze. The girls keep their hands on the reins, trading places and taking turns with the horses while others drink water or unwrap granola bars. I'm concerned about my horse, Joshua, wandering off, but Ron says, "He can't go far. There's an ocean on both sides." I relax. Ron's attitude is disarming as it is charming, prerequisites for working with people around large animals, I suppose. I learn that he grew up on the Navajo Reservation from the stories he tells and that he still organizes rides in northern Arizona. He's leading a group into Chaco Canyon in a few weeks.

Where the Anza Trail keeps us out of the hat-dislodging, eye-jabbing thickets, male vermilion flycatchers in courtship display fly straight up out of the mesquite trees, pumping impossibly tiny black wings as they corkscrew upward to impress females with their silly-looking antics. I laugh, watching them, while behind me Karen's horse suddenly decides he'd enjoy rolling in a sandy spot on the trail. Karen rides him to the ground like a rodeo queen and steps out of the saddle as if it were a practiced move.

We've never owned horses, but all the girls are practiced riders, both western style and English, experiences due mostly to the kindness of friends and relatives. My wife and daughters have ridden in Spain as well, although I've never been there. Today, they wear the boots they bought in Madrid.

Actually, I'm amazed at the opportunities my family has had. Jessica spent a semester of her junior year in Cork, Ireland, where she lived with a family and took courses in Gaelic language and Irish ecosystems. Because of her $10,000 annual Baird Scholarship and her other academic awards at the University of Arizona, she managed to save enough money to study abroad. She's studying wildlife biology and has become an accomplished photographer.

Kasondra's second trip to Europe lasted a month during the summer following her senior year. She earned it by saving her money and doing well in school. She was president of the National Honor Society at Pueblo High School, graduated seventh in her class, and also won scholarships to the University of Arizona, where she has just finished her sophomore year studying public health. She talks of joining the Peace Corps after graduation.

Melissa will take the same trip to Europe this summer.

All of this is possible only because of Karen's stubbornness. She focused her will on raising our daughters well, and the girls responded in kind. Education was primary. Karen would do whatever it took to launch the girls into college. Melissa will tell you, even now, that from her earliest memories, college was the goal for her life, over marriage, over having children. I followed Karen's lead. Both of us became proactive in the girls' education, from being a presence at their schools, writing letters and creating lesson plans for their teachers, to pouring what meager resources we had into the girls' interests—reading, science projects and fairs, girls' chorus, traveling, and horseback riding.

The Anza Trail ride, courtesy of a grant I received from the Arizona Commission on the Arts to write this book, is my gift to them to celebrate the end of another school year—Jessica's third in college, Kasondra's second, and particularly, Melissa's graduation from Pueblo High School, where just this past Tuesday she gave a speech to three hundred classmates as valedictorian. The girls have been looking forward to this adventure as much as I.

Peck Canyon, 69 River Miles

Five miles downriver from Rio Rico, I can see where Peck Canyon joins the Santa Cruz, sweeping in from the west where, near its headwaters eight miles away, it carves out a narrow, fire-walled channel between the Tumacácori and Atascosa mountains. This is Hells Gate, a place I had hiked to only two months ago. There, brick-red cliffs drip with black and sulfur-hued lichens, and some people see the face of the devil in the brimstone pinnacles.

The canyon is named for Artisan Leslie Peck, a miner and rancher who in 1886 lost his family in one of Geronimo's last raids. Peck's daughter-in-law, Marguerite O'Brien Peck, tells the story, discrepancies and all, as part of a family history she compiled with other stories over a period of thirty years (1937 to 1966). Her book, *In the Memory of a Man*, is a rambling, unpublished 500-plus-page tome, which Marguerite herself typed, single-spaced on legal-size paper.

Artisan Leslie, whom friends called "Al," met his wife, Petrita (or Beatriz) Quen, in Hermosillo, Sonora, where he worked in the Minas Prietas silver mines. In 1883, he gave up mining, packed up his wife and baby daughter, Mary, and bought a farm in Kansas with his father, but he "didn't like the arrangements and decide to come west again."

Apparently, his family didn't approve of his marriage and Petrita suffered their scorn.

Peck went back to the Prieta silver mines and "lived on coffee—the Mexican kind, beans, and tortillas." Then, sometime in 1885, his daughter Mary died from typhus or yellow fever. Peck took ten days off from the mines, traveled to Nogales, and bought a ranch in this canyon that now carries his name. "I paid a hundred dollars for a 'right of way' to the man who owned the shack and the ranch I wanted . . . up the canyon from where the Calabasas school is," he told Marguerite. There, he built a house "with a dirt floor," quit the mines again, and started buying cattle from Mexico, saying that "with good bulls and Mexican cows, a fellow can grow out and build up a herd."

But tragedy would follow him to this canyon as well. Marguerite says that "Dad Peck" never talked much about the Indian raid and she didn't press him for details because it clearly continued to upset him. Most of what she learned came from other sources.

On that day, April 27, 1886, Peck was working his cattle with his friend, Charley Owen. "We were about a mile from the house, and neither of us had any kind of a gun on us." When twenty or thirty Indians rode over a hill and began surrounding them, Charley shouted, "Indians, Peck! For God's sake, run!" Neither had a chance. Geronimo, Naiche (the son of Cochise), and his band of Chiricahua Apaches, who had only a month before bolted custody during a rainstorm after agreeing to surrender to General George Crook, quickly ran them down. They shot Charley from his horse and he died where he fell. Peck, his horse still tied to a two-year-old bull he had roped, had nowhere to go. The Apaches stripped both men, but for some reason left Peck alive.

Thomas Casanega, a former deputy sheriff who was living in Nogales at the time of the raid, would later describe this as "a mere mystery." "Peck must have acted like a crazy man," Casanega says in his unpublished reminiscences. "It was known in Arizona that Indians never attacked a crazy man. So, the Indians made him pull off his boots and told him to go bare-footed towards his home, which was two miles toward the Canyon."

Tubac historian Elizabeth Brownell, however, thinks what saved Peck may have been his red flannel long johns. "Possibly one of [the Indians] recognized him as a white man who had done them a favor in Mexico. When Peck was working in the Sonora mines he used to roll his shirtsleeves up over that same colorful underwear and the Indians, who liked him, called him Red Sleeves."

Peck stumbled with bleeding feet toward a black plume of smoke that twisted into the sky. He found his home collapsing in flames, its furnishings strewn across the ground. The Apaches had kidnapped Trinidad Verdin, Petrita's young cousin, who had been staying with the family. (A rancher in Mexico would rescue her six weeks later.) His pregnant wife and fourteen-month-old son lay dead.

The *Tombstone Epitaph*, reporting about the murders, claimed that Peck had said, "I will kill every Indian in this country," and then mounted his horse and rode off never to be seen again. The truth is less romantic but much more heroic. After the killings, Peck never returned to the canyon but nevertheless remained in the area with its painful memories. Moving to nearby Nogales, he married Carmen Cañez in January 1887 and started another family. The two produced a son, Arthur (Marguerite's future husband), and four daughters.

Over the years, Peck developed several mining interests, took a position as a superintendent of a mining company, and operated a livery stable on Morley Avenue. He became one of Nogales's leading citizens, serving on the city council and the county board of supervisors. His misfortune with Geronimo, his prominence as a public servant, even his friendship with people like Colonel William Greene never made him famous, but neither was the man entirely obscure. He was a man sculpted by this land, a place more raw and penetrating than most. His legacy remains today within the community he helped raise out of the rocky hills at our border with Mexico.

For decades I've driven past the highway sign on I-19 north of Nogales that marks the canyon named for the little-known miner and rancher. I've always wondered about the name. The massacre was unknown to me, although I'd written about Geronimo and his escape from General Crook on that stormy night just before the murders. I learned that Doug Cumming, whose family bought the ranch from Joe Piskorski after Peck sold it to him, had placed a marker commemorating the site.

After a couple of failed attempts at locating the marker, I called Dan Koskuba, a retired high school teacher and local history buff who lives in Rio Rico, and he agreed to take Jessica and me to the site. "I'm working on a book about Al Peck, and I've been looking for an excuse to get up there again," he told me on the phone. We met on a Friday morning at Garrett's Market off Exit 17, and Dan drove us through a maze of streets to the forest roads leading deep into Peck Canyon, his green 4x4

Cherokee Scout deftly maneuvering over the rutted and rock-studded tracks. When the Scout could go no farther, we walked.

"Ever been hit in the face by a Texas gate?" Dan asked me, referring to all the fence gates I had to open so he could drive through them. The gates lock in place with a chain attached to a heavy stick, which, if you're not careful, can spring loose and swing into your teeth like a pair of whirling nunchakus. Dan, his bearded face shaded by a blue "Hoboken" ball cap (he's originally from New Jersey), gripped a heavy walking stick, protection from rattlesnakes and falling, he said.

Jessica and I shed our sweatshirts as our skin heated up. She wore her hair loose, her blonde curls unfettered by the usual colorful scrunchie. Her camera, her only burden since I carried our pack and water, she cradled in both hands as she stepped in the gauzy sunlight.

"We're close," Dan said. "See that mesa up ahead? That's how I found it the first time." I stopped to make a few notes, and Jessica and Dan moved out of sight. Soon, I heard their excited voices. They had found the place.

I noticed dark bricks scattered on the ground, which sloped away sharply toward a broad drainage dotted with walnut and cottonwood trees. Nearby, a column of cemented native rock rose three feet among the mesquite and catclaw. On its top, a tarnished metal plaque read:

MRS. ARTEMUS PECK
and her infant
were killed here by
Apaches Apr. 27, 1886
Peck home site 200 ft.

I remember searching among the bricks for something less tangible than lichen-darkened, fired adobe. I wanted the tragedy to make sense, to seem less like a random act of violence buried under a hundred and twenty years of dust and gravity in some remote desert canyon. How could it have happened? Why? With so many variables coming together in time and space—Geronimo's escape and the location of the ranch among them—all of them colliding at this place, it seemed as though the whole universe were arrayed against one man. As if God, running out of Jobs, went looking for someone else to bring too much attention to.

The questions still remain on my lips. What was the desolation I felt there, as sharp and gripping as catclaw spines?

Monument at Peck Canyon

I watched Jessica take photographs, her body small against the span of canyon and mountain. There was nothing swift about her movements except the flash of her eyes. An artist at work, measuring the frame, composing her subject. I listened to her boots on the stones, to the muted lashes of branches against her clothing. I felt her as she stepped around me, father and daughter, our shadows falling together as water poured out. How would it be, I thought, to lose a child?

Dan, who had eyes for artifacts, showed me a bullet casing. "Smith and Wesson .38," he said. "I'd have to check, but it could be contemporary." He found chips of blue china, horseshoe nails, pottery sherds, and a single broken diaper pin. Then he handed me a pearl button that looked like it came from a child's nightshirt. "Maybe it belonged to the baby," he mused.

Although Geronimo's fame soon grew to legendary status, his warrior days would come to an end within months of the massacre. He surrendered to General Nelson Miles, General Crook's replacement, in early September at Skeleton Canyon, boarded a train bound for Florida, and never returned to Arizona territory.

Peck's legacy, however, the one thing that does make sense of his tragic life, remains today. As Jessica, Dan, and I hiked out of the canyon, I considered how diminished I felt in that place. Compared to Al Peck, I was an infant in that landscape, foolish and uncomprehending. How could I even begin to touch the world he knew without panicking and fleeing? Peck loved his family, and for many long years he lived with the graves of those he loved, remembering dates of marriage and birth, recalling in some personal ritual the days and events that could have been but weren't.

In my pack, I carried a photo of Artisan Leslie Peck, taken some years before he died in Nogales in 1939 at the age of ninety-four. It showed a face wrinkled with age but still framed by a full head of white hair. His large ears, thick, pursed lips, and broad nose couldn't match the enormity of his gaze.

My family and I ride under a giant walnut tree, the first I've seen along the Anza Trail this side of the mission. I point it out to Bill and Dick, who ride just ahead of me, saying, "This is the tree Nogales is named for."

"Is that right?" Bill McClain says.

"Yep," Dick agrees. "Nogales means 'walnut.'"

"It was first named Border City," I add. "And then later Nogales for the walnut trees that once grew there. The scientific name for walnut is *Juglans*, or *Jovis glans*, so I like to tell people that Nogales is the town named for Jupiter's nuts. In the Golden Age, it is said, men lived upon acorns, but the gods lived upon walnuts."

Bill McClain is Ron's partner and the comedian of the trip. He keeps an eye on Karen in case her horse decides to take another sand bath. Ron thinks they should trade horses. Karen tells Bill about Melissa's acceptance to Harvard and Bill relates another of his corny jokes. He says in a wonderful western twang accented through thick gray sideburns and mustache: "There's not much difference'n college folks and me. They went to Penn State. I went to the State Pen." Jeez, I think. How much does he know about us?

I look behind me to Melissa, who's smiling, but I know her mind is awhirl. She has recently come through a difficult, heart-wrenching time of decision. A month ago, Melissa and I visited all the California universities she'd applied to, some of which had already accepted her: UC Berkeley, UC Santa Cruz, Stanford, and UC Santa Barbara. All were great schools, and some even had beaches, which seemed to be one of Melissa's prerequisites, but money was the primary issue. After we returned to Tucson, Melissa learned that Harvard had also accepted her and that the school was offering a scholarship of $42,000 a year. And then a letter arrived FedEx from Arizona's Flinn Foundation. She'd been chosen from among four hundred applicants for the prestigious Flinn Scholarship—a full ride to any one of Arizona's three universities, plus funds for housing, living expenses, and summer travel. These two awards effectively put the California schools out of the running, but now she had to decide between Harvard and the Flinn. It took her weeks. Weeks of research and making contacts, weighing pros and cons, talking to friends and teachers and counselors, Harvard graduates and students, Flinn Scholars present and past. Still, she couldn't decide, despite an approaching deadline. We flipped coins (which alternated perfectly between one choice and the other). We looked for signs in the heavens. Melissa's heart wanted Harvard, but her head said take the Flinn.

In the end it wasn't the money that made up her mind but her friendship with a Hispanic girl, the high school salutatorian and the first in her family to go to college. She would be Melissa's roommate. On the night Melissa made her choice, she laid out on her bedroom floor all of her pictures of Harvard, turned out the lights, and cried herself to sleep. She would stay in Tucson and enroll at the University of Arizona.

Melissa, my home girl. My river girl. She could go anywhere her heart desires but she chooses to remain in the desert of her birth.

She rides behind me now, followed by her sisters Kasondra and Jessica. I couldn't be more proud of all three of them. Despite the challenges, my daughters have already made something of themselves.

Tumacácori Highlands, 69 River Miles

From the Santa Cruz River where we ride, to Hells Gate in the Tumacácori Mountains on our western horizon, Peck Canyon wriggles twelve miles of canyonland out of eight miles of terrain. This is tough foot country. The sinuous bottomland of the corridor works you over, weakening leg muscles as your steps shorten in the loose sand and your route doubles back over and over. Thorny brush grabs at you with malicious intentions. Your hat and sunscreen may or may not protect you. This is one place where it's easier getting into Hell from the high ground.

Two months ago, in early April, I hiked to Hells Gate from the high western flanks of the Tumacácori Mountains after enduring a kidney-punching ride in the back of a pickup truck with a group of artists and scientists brought together by Sky Island Alliance. The nonprofit organization promotes the preservation and restoration of native wildlife in the sky island region of the Southwest and Mexico. On this particular occasion, Sky Island had asked us to help in its campaign to designate the seventy-six-thousand-acre Tumacácori Highlands as a wilderness area. Mike Quigley, the campaign coordinator, called the event "Art in Wilderness" and had invited more than a dozen poets, musicians and songwriters, filmmakers, and photographers to join the alliance's staff for a long weekend of camping in jaguar country. Mike said I could come if I agreed to set up Dick Shelton's tent.

When I first met him in 1989, Dick was writing *Going Back to Bisbee*, a book about his travels across southeast Arizona in a Volkswagen microbus he called Blue Boy. He would often read me passages from the manuscript, and I remember becoming lost in his stories of fat, home-invading toads and squirrel-swallowing snakes, and laughing at others, like the time he took a group of people bird-watching and found "rosy-breasted pushovers" and "extra-marital larks." Dick became my inspiration, my mentor. I wanted to write like he writes. At a time when it seemed the universe was arrayed against me, his creative writing program kept the gravity of my life from crushing me. He taught me about metaphor and the rhythm of words. I learned the language of poetry instead of the language of despair.

Trevor Hare dropped us, rattled and shaken, at Corral Nuevo, our back door to Hells Gate. Trevor, a conservation biologist with Sky Island, reminded me of a young Edward Abbey when, the day before, I shook his hand and his voice boomed in greeting through his thick mustache and beard. I instantly felt welcome.

At Corral Nuevo, we started for the pooled water of upper Peck Canyon. An orange-headed bombardier beetle tumbled through the grasses, holding its explosive blue-gray hindquarters in check as I attempted to lift it onto my hand. The tracks of other larger passersby pocked the sodden ground where grasses gave way to mud. We identified raccoons and skunks and the large cloverleaf of a carnivore, which my imagination enlarged into a heavy feline paw. I could see it in my mind, just before dawn, a shadow that slides to the water's edge and stills, the huge head and level amber eyes lowering to barely break the surface tension with curled tongue. It is an image that has haunted me since the day I first learned that jaguars are returning to this wild borderland country.

I have been fascinated by jaguars since I saw my first one when I was ten years old. Although caged, the animal seemed alert, spring-loaded. This was a creature of mountain ridges and rugged side-canyons. I felt that it knew no boundaries, that its enclosure was superfluous. Later, as I began to develop a style with colored pencils, I drew that scene of the jaguar with its head lowered and eyes fixed straight ahead. It wasn't my best, but the eyes are exactly as I remember them, partially lidded and penetrating. Those twin yellow points still look right through me. How was it that from a sketched initiation with that magnificent animal I would find myself in this wild place?

Jaguars have stalked the imaginations of people for thousands of years wherever the cats and humans have crossed paths. In southern Mexico at San Lorenzo Tenochtitlán, a two-thousand-year-old nine-foot Olmec head of carved basalt bears the paws of a jaguar across both colossal temples. *Olmec* is an Aztec word translated one way as "Mouth of the Jaguar." The Olmec was the earliest organized society in Mesoamerica, and jaguars featured prominently in its mythology and art. One stone image thought to best illustrate the relationship between the Olmec and the jaguar, one that graphically portrays Olmec mythology concerning a race of half-human, half-animal "were-jaguars," shows a jaguar mating with a human female.

If the Olmecs had jaguars on the brain, the Maya, who followed them, took the symbolism to the level of obsession. The ruins of their cities on the Yucatán Peninsula, like those at Balamku and Chichén Itzá, depict elaborate jaguar imagery, some with the cats painted red and encrusted with jade and turquoise or with carvings of human hearts in their claws. The Maya both admired and feared the jaguar. The cat was a central figure in their stories, rituals, and sacrifices and embodied beauty and strength, cunning and mystery. Spread the jaguar's skin, say the Maya, and you spread the heavens of a starry night.

I thought about how the Maya revered the jaguar as we followed the drainage with its ellipsis of dark pools so deep they held the sky. With clouds in the rocks, we entered the heart of the Tumacácori Highlands,

Tumacácori Highlands

new domain of the mythical jaguar. The dappled cat, third largest in the world and the only one in the New World that truly roars.

Everywhere, I saw spots. Even the rocks were mottled, dimpled and pitted with smooth water-carved holes. One rock the size of a house had a dozen round pocks in its face, the giant boulder cleaved in two after some cataclysm had rolled it over to settle in this quiet repose.

Farther on, we discovered geodes, whole great seams of geodes still welded in original rock. Their broken-egg shapes and hollow, crystal-lined interiors mesmerized one songwriter I had been hiking with, dark-eyed Cantrell, and she searched for a few stone-loosened prizes to share with us. Cantrell, whose name means "Little Singer," found something musical in their quality perhaps, but all I saw in the elongated and embedded geode seams was jaguar skin.

Cantrell Maryott, a third-generation Arizonan and native Tucsonan, creates songs of sacred places in what she describes as an "indigenous Sonoran Desert voice." She calls it "chanteling," a harmonizing within acoustically charged spaces, music influenced by landscape with all its unique weather, flora and fauna, history, and human culture. "It is as much about the place as it is the song," she says. "The space between notes, the stillness between heartbeats."

Wilderness exists in the stillness between heartbeats, I think, remembering a question posed by Terry Tempest Williams, my favorite nature writer: "Who can withstand the recondite wisdom and sonorous silence of wilderness?" This is the question. It's like asking, Who can withstand the wolf or jaguar? Or the teeming masses that plough through soil and sky and skin? Or maybe the real question is, Who can *stand* it? Can we stand wilderness for the sake of wilderness?

The Tumacácori Highlands nurtures communities of woodlands knit together with riparian drainages raising giant canopies of ash and cottonwood. Here, Mexican free-tailed bats migrate in spring, millions pluming as smoke out of forgotten mine shafts to separate moths from their wings in frenzied nighttime feeding. Another Mexican migrant, the lesser long-nosed bat, appears just as the yucca begin spiking up blooms in northbound waves, the bats' pollen-wet faces a perfect inversion of the moonlit flowers.

Adjoining the Tumacácori Highlands, the 7500-acre Pajarita Wilderness sustains Sycamore Creek, whose perennial waters seep southward into Mexico and create a haven for tropical birds like rufous-capped warblers, rose-throated becards, and the birdwatcher's prize, elegant trogons. This narrow, twisting funnel is where north really meets south, geographically and specifically. Virginia creeper, yellow columbine, and Utah serviceberry range no farther south than this place, while tropical species such as pineapple-like ball mosses, the only bromeliad found in Arizona, Chihuahuan hook-nosed snakes, green rat snakes, brown vine snakes, and barking frogs creep north from Mexico. In these warm tannin-dark pools, a desert fish called the Sonora chub enters the United States only here to swim among dead leaves, algae-furred rocks, and muddy jaguar tracks.

The day we arrived to join Sky Island Alliance, Dick Shelton and I hiked from our campsite at Yanks Tank to the Sycamore Canyon drainage, passing markers indicating we had entered the Pajarita wilderness area. We clambered over rocks and around pools for three hours, spooking lime-colored leopard frogs and sleek, dark Sonora chub.

On our return to camp, Dick reminded me that that day, Friday, April 8, a partial solar eclipse would occur in the afternoon. I poked a hole in a page of my notebook to create a pencil-width lensed image of the sun on a second page. The sun still held to its orb. Then I recalled a Maya myth that says the end of the world will come when jaguars rise from the underworld to eat the sun and the moon. An eclipse will foreshadow the event. Friday was the day of the new moon. Somewhere near New Zealand at sunrise, an annular (ringlike) eclipse had begun sliding eastward into a path of totality that would streak across the South Pacific toward Costa Rica, slicing through the heart of jaguar country where the sun would be devoured just before it set.

Back at camp, people wanted to know the state of the eclipse, so I pulled out my pinhole viewer to check on the condition of the sun.

We entered Hells Gate, heart of the proposed wilderness area, as a western sun seasoned the canyon walls with paprika and cayenne. Layered rock rose skyward on both sides where Peck Canyon slashed obliquely, cutting through one bulwark only to angle back and drive through another. This had been a long war fought with gravity and water and time against a fortress of stone.

Cantrell and I had outpaced our group, and she immediately climbed to an alcove of rock to our right and leaned with her hands into the stone as if praying. And then her voice began to fill Hells Gate with the body of its sound, a kind of lithic harmony that came from nowhere and yet everywhere and connected both of us to the center of that world. She was chanteling. She sang a song that could have arisen from an ancient Maya woman atop her temple at Chichén Itzá. And she sang it with the canyon's own throat.

That evening at the campfire, our last night in the Tumacácori Highlands, smoke sifted through us like incense. We ate from plates heavy with beans and tortillas. After dinner, we sat in a circle while Sergio Avila, a Mexican wildlife biologist working with Sky Island Alliance, spoke about his work with jaguars. Sergio had lived with the indigenous Tarahumara people in the Sierra Madre while learning to track and monitor the animals. More recently, he had joined researcher Jack Childs in setting up and operating camera traps to document the presence of jaguars in Arizona. Sergio passed around these photographs, telling us about the day he picked them up from the film developer. "I was flipping through them, seeing deer and raccoons and the usual nothing. And then there it was. A jaguar! I had tears running down my face."

Sergio then produced a plaster cast of a jaguar footprint. It was huge, the size of my palm and each of the four toes like fat thumbprints. I held the jaguar cast in my hand and asked him about jaguars in the area.

"We've photographed two different males," he said. "And we believe there are females, too." I peppered Sergio with more questions, but others wanted him to make the sound of a jaguar.

"Once caught," he warned, "a jaguar won't cower at the end of the tether. They come straight for you. This is the sound they make." Sergio lifted a white plastic bucket and stuck his head inside. Then, he began to purr in a rapid, punctuated staccato, with each throaty trill growing in volume and depth until the hair on the back of my neck began to rise as the woods filled with the sound of it.

Late that night, I walked alone to my tent under a pelt of stars. Jaguar heavens. Somewhere in the moonless darkness I heard sounds, a thumping in the grass beneath the oaks. Sergio's jaguar roar rushed inside my head. The animal was no longer mythical to me. It was as real as the day before when I had taken out my pinhole viewer to check on the sun and there, along its smooth arc, was a perfect bite mark.

Mission Tumacácori, 74 River Miles

We break for sandwiches at one in the afternoon inside a branch-tangled mesquite bosque a few miles from the Tumacácori mission. Birds call from the undergrowth and insects zip through the air while the humidity drowns us under its heavy wet breath. I watch my girls set a picket line for the horses and tend to their mounts with soft hands and kind words. Kasondra asks if her horse can have an apple as we retrieve our lunch from our saddlebags. Where I sit on the ground, the earth smells like toads.

In places, the ground is turned inside out, disturbed by burrowing animals. I think about all the creatures that might plow the ground into wide tunnels with their fresh mounds of cinnamon dirt, hoping for the more exotic and sexy like badgers or box turtles rather than something ordinary like gophers or ground squirrels. Kasondra, always my pet-enthusiastic child, has a thing for box turtles, as well as fat-lipped cichlids (oscars), zebra-skinned salamanders, bubble-headed spadefoots, and one six-foot green iguana, all of which have lived in her room. But box turtles are one of her favorites of recent years. I should check her saddle bags, I consider, since this part of the Santa Cruz River adjacent to the Altar Valley is the westernmost range of the ornate, tan-on-straw, hinged-shell terrapins.

"My first box turtle's name was Herman," I told Kasondra, after a friend living in the Altar Valley gave her one after her tenth birthday. The animal had been living for several years in a small glass aquarium away from natural light and fresh air. If turtles can look anemic, this one certainly did. Its domed carapace and flattened plastron were dingy gray and flaxen instead of the normal olive and yellow; its flesh, when visible and not tucked into shell, seemed as pasty as doll skin. This was an animal that needed the desert underfoot and sunlight overhead, not life in a fish tank.

I went to work constructing an outdoor enclosure in our yard. Deciding to make the enclosure oval in shape, I set about digging up a bathtub-sized area with pick and shovel. Kasondra helped by collecting rocks and suitable plants. I needed to go at least three feet deep if we wanted the turtle to have room enough to tunnel, especially before winter when I knew the reptile would disappear underground for a period of aestivation, the cold-blooded version of mammalian hibernation.

Kasondra wanted her box turtle, a yellow-eyed female she named Hermit, to be able to aestivate because she had read that the animals need this period of sleep to reproduce successfully. She had plans on finding a "husband" for Hermit.

It took all day to excavate the turtle pen, spread the pit with heavy galvanized fencing, stake up the perimeter with the same wire, and fill it back in. Kasondra constructed a "turtle house" of stones and wood that looked like a miniature open-sided Navajo hogan and planted a Texas ranger and fountain grass for shade. She decorated the enclosure to give Hermit "something interesting to look at."

Hermit accepted her new home with much grace, promptly burying the rocks and digging up the plants.

Hermit's color and attitude soon improved. She enjoyed being outdoors, basking in the sun and roaming around her enclosure to inspect every nook, every new insect that stumbled in. She developed a fondness for grubs and worms, and she would eat them with obvious relish on her face. The turtle had a lot of character for a reptile. What Hermit didn't like, however, was the new husband.

When Kasondra and I found Kermit, who was slightly larger and darker than Hermit, my daughter thought he would be a good match. I was suspicious. There was something in Kermit's eyes, something behind the fiery red irises.

Then Kasondra introduced him to Hermit, and we had no doubt of his maleness. His gonads were undoubtedly pumping quantities of testosterone into his hot reptilian bloodstream, too hot for any normal red-blooded male turtle. Kasondra was about to get her first lesson about sex, and not the kind of sex she already understood: that it takes both mommies and daddies to make babies. This was worse. This was turtle sex in all its glory. Before now, even I had been curious. How do turtles, shells and all, actually "do it"? Snakes at least can entwine. Lizards have arms and legs unencumbered with interlocking scutes of bone and keratin. How do turtles get around those horny breastplates? I had read about "the member" that looks like a purple flower.

I had to cover Kasondra's eyes.

Hermit, apparently, didn't want to be married to Kermit. She took one look at him with his red eyes bearing down on her and she took off for the fence line. There are times when turtles can be *really* fast. But Kermit would not be denied. He acted as if he'd never seen the female form—as it pertains to oblate turtle forms—before meeting Hermit. Until he finally caught up with her.

This is where I had to cover *my* eyes.

It turns out, I learned, that male box turtles have special adaptations for dealing with cumbersome shells when mating. One of them is extremely long—the tail. This very articulate, nearly prehensile appendage could pry open the valves of a modest and tight-lipped oyster. The other adaptation is a set of short, thick, and curved hind claws. These work to hold onto the female in a way that is most surprising, at least it was to me when I first witnessed Kermit employ them with Hermit. For once he made connection, his tail firmly in place between her plastron and carapace and his toenails locked tightly onto the rear margin of her shell, he would relax and flip over onto his back allowing Hermit to drag him around the enclosure as if she were pulling a toboggan.

I remember how Kasondra, less shocked by Kermit's display than I, gave Karen the play by play from the yard to the house. "He's going straight for her," she shouted to my wife. "She's trying to get away, but he's shoving and biting her. Now he's circling around her and climbing on her. He's turned over on his back! He has a big smile on his face!"

All that summer and into the fall, right up to the first Pacific cold fronts that carried the signals for aestivation, Kermit couldn't leave Hermit alone. Kasondra, instead of borrowing the name of a sexless Muppet character, should have bestowed on him the more appropriate name of "Randy."

After lunch no one wants to move. Karen and our girls relax on a plush carpet of mustards with Dick, Ron, and Bill adding the complementary arc to our rough semicircle. I notice that both Ron and Bill wear black cowboy hats and that Ron favors Double-H boots. Their shirts have long sleeves, unlike mine which doesn't have sleeves at all. But then, I'm not using a machete to open the trail before us, so the flesh of my arms is relatively safe. A large silver and turquoise belt buckle fills Bill's midriff. He could eat a T-bone off of the silver platter.

The girls are interested in Ron's horseback-riding business, which he says he's moving from Tucson to the town of Sonoita where he'll have easier access to the Santa Rita Mountains, one of his favorite places to take people. "You should see it," he says. "I'll take you all and you can write about it. It's beautiful." Dick agrees. He's already written about it.

Bill is a recent addition to the program and continues to run his own separate guide business for private hunting parties seeking game like javelina, quail, and deer. "I'd do this kind of work for nothing," he says, "but I have to support all my ex-wives." I laugh. "It's true!" he adds, a claim he makes at the end of all his tales.

I might imagine Ron Izzo as the Juan Bautista of our expedition, but "Wild Bill" is certainly no Pedro Font.

The Franciscan Father Pedro Font would accompany the Anza expedition to California and back, serving as chaplain for the colonists with "the added duty of observing the latitudes and directions on the way." He joined Anza on the second of August, 1775, at the presidio of San Miguel de Horcasitas, then the capital of Sonora, where Anza was assembling and training colonists he had previously recruited in villages from Culiacán to Alamos.

From his journals, I learn that the monk arrived almost two months early but occupied himself with visiting the mission at Ures, taking latitudes of the presidio, and suffering from an intestinal illness that would plague him throughout the journey. His writing is wonderful—among all the dry facts of his adventure, the dates and holy days, the details of his destinations and distances traveled, the lists of supplies and number of colonists—are occasional bouts of language where the padre's inner feelings begin to leak from his pen. One of these bouts occurs very early in his journal and concerns Anza's proprietary control over a piece of equipment: "The instrument with which I made the observations was an astronomical quadrant sent by the Viceroy to be delivered to me, but Captain Anza kept it in his possession, not wishing to deliver it to me, and only when it was necessary to make an observation did he open the box in which it came encased." In his own diary, Anza writes that the viceroy "provided the expedition" with the instrument, without even mentioning Font. The padre would continue to refer to the quadrant as "carried by the commander" among other complaints directed at Anza, such as his unanswered requests for a personal servant. It seems the two didn't get off to a good start.

Finally, on September 29, the party of soldiers, muleteers, and settlers numbering 177 men, women, and children had assembled at Horcasitas. Father Pedro Font sang Mass for their success and afterward gave a speech "exhorting everybody to have patience in the hardships of the journey" and to set a good example "for the heathen, as a mark of Christianity, without scandalizing them in any way." I'm guessing that the padre was a bit long-winded. The expedition left Horcasitas at four-thirty in the afternoon and made its first camp shortly after five, having traveled only a few miles out of town.

Sixteen days and 150 miles later, after several episodes of lost mules and scattered and lost gear, the expedition arrived safely at Tubac.

It was here that Pedro Font took his side trip to Calabasas to say Mass for its residents. Afterward, the Franciscan detoured again on his way to meeting up with the colonists at Tubac, apparently uninterested in helping with the preparations for the next leg of their journey. In his diary on October 15, 1775, he writes: "I stopped at the mission of Tumacácori, which is on the road a league before reaching the presidio. Here I found fathers Fray Francisco Garcés and Fray Thomas Eixarch, who were to come with the expedition to remain at the Colorado River, and here I remained with them . . . while the expedition halted at Tubac."

The padre would spend the next four days in bed at the mission's *convento*, or priests' quarters, with what he described as "my illness with flux."

A mile and a half from the mission at Tumacácori, we dismount in a wide drainage coming out of the northwest that I believe to be Josephine Canyon. We lead our horses down its stony, shrub-choked floor. Thin wands of seep willow in tight stands erase any trail, and our route only grows more impenetrable as we close in on the Santa Cruz River. We halt outside the bosque while Ron and Bill go on ahead with machetes to clear our path. Twenty minutes later, we're leading our horses again.

Where the mesquite fretwork gives way to cottonwoods and willow, we mount our horses and ride under an open canopy of malachite leaves.

The gallery darkens, but not because of the trees. Thunderheads, which have been building all day, now hide the sun. I can smell wet dust as the wind stirs and lightning flashes and fat cold drops begin ticking against the hard leaves. Karen and the girls don ponchos, which the wranglers pull from saddlebags. They wear them too, but Dick and I prefer the rain on our skin.

When we cross a dirt road, Ron points to the west along it and tells us that Tumacácori is not far away. "We'll stay on the trail just this side of the mission grounds."

We continue on the east side of the Santa Cruz, which is actually thought to be the site of the original mission, the place where the native people prepared three mesquite-brush ramadas in anticipation of Father Kino's arrival. The beautiful Franciscan church we see today did not even exist in Anza's day. From the time that Father Kino established the mission in 1691 until the brown-robed monks built this present church between 1800 and 1822, Tumacácori was mostly simple structures and adobe dwellings.

Kino called it San Cayetano de Tumacácori, an O'odham village of forty or so dome-shaped houses made from bent saplings covered with brush and earth that dotted this bank of the Santa Cruz River. (*Tumacácori* is an O'odham word probably meaning "rocky flat place.") The priests celebrated Mass, performed baptisms and marriages, slept and prepared meals in a plain adobe house, but for the first half of the 1700s lived at Guevavi, the *cabecera* where they had built their church. After the Pima Revolt of 1751, the mission moved to the west side of the river and became San Jose de Tumacácori. What we begin to recognize as a church, the Jesuits completed there in 1757, where it served the community through disease, Apache raids, and political upheaval for the next sixty-five years.

When King Charles III banished the Jesuits from all Spanish lands on February 27, 1767, the mandate reached the Marques de Croix, viceroy of New Spain, on June 24 and then Juan Claudio de Pineda, the governor of Sonora, on July 11. On July 25, the order came to the captain of the garrison at Tubac—Juan Bautista de Anza. I can only imagine how this decree must have pained him. Anza came from a family of devout Roman Catholics. His father assisted the Jesuits in establishing their new missions. His first wife's brother, Joseph Manuel Diaz del Carpio, was a priest (although fortunately not a Jesuit) whom Anza had persuaded to be Tubac's chaplain. Furthermore, Basques—Anza's people—had always been supportive of the Society of Jesus because the founder of the religious order, Ignatius Loyola, was a Basque.

Perhaps it was the teachings of Ignatius Loyola, particularly his insistence on obedience to the Pope, that gave reason for King Charles III to expel the Jesuits. Spain was already pulling away from Rome and moving toward a more secular government, and the Jesuits were a powerful order with a foreign allegiance. Or perhaps the missionary system under the Jesuits was too costly and Spain wanted her possessions to become more profitable. Whatever the reasons behind the decision, or the emotional weight of its consequences, Anza was a soldier with a duty. The captain would carry out the order: to arrest all Jesuits serving the missions on the Sonora River.

Ten days later, Anza wrote a letter to the governor of Sonora saying that he had almost completed his commission, that three Jesuits were already en route to Mátape for deportation, and that he "would have already gone on the march before now but for the illness of Padre Perera." The seventy-one-year-old Father Nicolas de Perera could not ride a horse. So Anza, giving him time to recuperate, had a reed stretcher made for him that would allow him to be carried on pack mules. Unfortunately, even with all of Anza's care, the padre died before leaving the country.

With the Jesuits in exile, the Franciscans now arrived on the scene. In 1768, the lanky, curly-bearded, hair-shirted Juan Crisóstomo Gil de Bernabé, the first Franciscan priest assigned to Guevavi, decided he didn't like the living arrangements and moved himself and the cabecera to Tumacácori. For the first time, Tumacácori had a resident priest.

The Franciscans who followed him there, Fathers Agorreta, Ximeno, Zuñiga, Clement, Moreno, and others, set about redecorating the church, constructing dwellings, and building a wall around the mission. Tumacácori began to take shape.

Around 1800, a blue-eyed, ruddy-skinned Franciscan named Fray Narciso Gutierrez started construction of the present church, replacing the modest Jesuit structure. The original design was ambitious, maybe too optimistic. Father Gutierrez staked out the foundations fifty feet behind the Jesuit church, orienting its one-hundred-foot length north-south. The padre wouldn't be outdone by the frontier baroque beauty of the mission at San Xavier del Bac. He, too, would build his church in the shape of a cross with a barrel-vaulted nave and transepts. A dome would rise at the center of the cross and another over a bell tower. Spanish and native laborers, dispatched to the nearby Santa Rita Mountains, would cut timber for supports and gather limestone for plastering the walls.

Others would dig soil to mix with water and straw, forming the mud into bricks to bake in ovens or dry in the sun.

But after the workers laid a five-foot-wide, three-foot-deep cobblestone foundation of giant river boulders set in mud mortar, the financing began to evaporate. It took years before the adobe walls stood at seven feet, a third of their intended height. It became clear to Gutierrez and the other Franciscans that they would have to rein in their vision. The final plan, which would take over twenty years to finish, involved moving the central dome over the sanctuary, eliminating the transepts, and replacing the barrel-vaulting with a flat roof. The bell tower remained in the plan but never reached completion.

Yet the result was still astounding. In the decades after 1822, if you were riding north along this section of the Santa Cruz River, your first glimpse that something magnificent lay ahead would be red walls and a brilliant white dome towering out of the green mesquite trees. Imagine the scene. As you draw close, you might notice scaffolding circling an unfinished bell tower and walls with arched gateways enclosing rooms and courtyards, fountains and gardens. People are coming and going, men and women and children in brightly colored clothes, some carrying woven baskets of grain or clay ollas of water. A robed friar on horseback scatters chickens and dogs in his wake.

As you approach the front of the church, its colorful detail strikes you: a façade of yellow behind four pairs of stacked columns painted red, their Egyptian-style capitals in yellow and black. Deep niches, one between each column pair, show blue behind their wood-carved and painted statues of saints. When you enter through the arched doorway, the outside world's light and sounds ebb. You smell the wax of dozens of votive candles; their heat is a curtain you must part to pass through. Your dusty boots suddenly seem profane as you realize you are standing in a holy place.

People stand or kneel along the seventy-five-foot rectangular hall, or nave, which has no pews. Images fill your mind as you walk past the baptistry on your right and out from under the choir loft and continue down the nave though oblique pillars of sunlight. Along the walls, four side altars hold smoking candles under more niches where Mexican baroque statues stare at you with their dark, sad eyes: perhaps St. Francisco, patron of Father Kino, or San Cayetano, patron of Calabasas. The gaze of St. Bonaventure or St. Peter of Alcántara might cause you to pause. Or a young married woman, kneeling beneath the image of St. Anthony of Padua, praying to be blessed with a child. Other saints and apostles in colorful paintings, carvings depicting the stations of the cross, and imagery of the Virgin Mary cover the walls among exquisite stenciled designs of Catholic symbolism.

Straight ahead, steps carry you up to the center of the sanctuary where tall candlesticks adorn an altar. Beneath the high-domed ceiling, an ornately carved angel watches over the pious with flared wings.

If services were at daybreak, worshippers would crowd the nave. A priest in golden-hemmed purple or green or white vestments would enter the sanctuary from the sacristy on the right to celebrate the Misa Mayor, or High Mass, which he sings in Latin. The choir would respond with Gregorian chant and songs, the native boys and girls, accompanied by musicians playing trumpets, oboes, flutes, and guitars, filling the church with a harmony of instruments and voices. For those present, the place and its celebration would seem like the glory of heaven.

For the people of Tumacácori, however, heaven on earth would last only a few years. In 1828, after Mexico gained her independence, the mission lost her last resident priest when the Mexican government ordered all the Spanish out of the country. Although native Mexican priests still visited, and settlers and the O'odham people continued living there, years of drought, political strife, and Apache attacks finally forced these remnant people to abandon the mission by 1848, severing a continuous thread of community first noted by Father Kino 157 years earlier.

Karen and I visited Tumacácori last September when the mission's many hummingbird feeders were ablur with the energetic birds. We'd never seen so many hummingbirds in one place, so many tiny whirring angel wings.

At the convento, I thought about Pedro Font recovering from the "flux," although the mud-walled rectangular roofless structure is probably a later construction than the convento of Anza's time. The six- and eight-foot adobe walls, chocked with thumb-sized rocks, still held a mosaic of dirty plaster on the interior; a single clay pot rested in the mouth of a fireplace. Little more than an outline of two-foot melted walls marked the place of the original Jesuit church, which lay to the north of the convento.

Karen and I continued our circuit of the mission grounds, passing the remains of irrigation ditches that once carried water for two miles to vegetable gardens, vineyards, and orchards. At the cemetery, we walked around the circular and unfinished mortuary chapel, knowing from our

Tumacácori mission

tour brochure that more than six hundred graves lay beneath our feet. The ground practically heaved with the dead, victims of Apache raids, disease, and starvation among other causes related to life in a land where living and dying was a crapshoot and the dice were weighted toward snake eyes on the come-out roll.

Inside the Franciscan church, Karen and I stood at the altar steps above where the energetic church-builder Narciso Gutierrez and another priest, Balthazar Carrillo, were buried until 1935 when their bodies were moved to the mortuary chapel at San Xavier del Bac. (Gutierrez never saw his church completed, but he would lie beneath the sanctuary for 112 years.) Gregorian chant issued from speakers hidden somewhere near the missing choir loft behind us. Thick walls rose high above our heads, the plaster now broken and pitted with great sections torn away, exposing a soft mocha-colored adobe beneath it. The nave's altars were crumbling pediments beneath their ruined niches, and tunneling animals had left their marks during the years that the mission stood alone and silent. In other places, however, white plaster still showed the stenciled colors and designs—scrolls and flowers, red draperies and dark garlands—that hinted at the church's former magnificence, especially under the high brilliant dome. Standing there with my neck craning heavenward, I thought it was a miracle that after two hundred years entropy hadn't brought the dome to the ground.

After the Treaty of Guadalupe Hidalgo was signed in 1848 and the communities at Tubac and Tumacácori suffered yet another decline, the mission would wait another sixty years for its redemption. Where the church no longer provided a place to meet people's spiritual needs, it now offered a place for physical needs. Settlers, looking for building material, removed the roof's timbers. Travelers found rest and warmth within the thick walls of the sacristy. The soot from their campfires on its ceiling testifies to the visits of forty-niners and soldiers and ranch hands as the plaster of its walls records their scratched names—"Sutton 1919," "L. M. Lemon, Phoenix 1890." Even John J. Pershing, a U.S. Army general on patrol along the border, left his signature.

Then, in 1908, President Theodore Roosevelt established Tumacácori as a national monument. The National Park Service replaced the roof in 1921 and again in 1947 and 1978 and restored other portions of the mission church, like its tiled floors, which treasure seekers destroyed while searching for mythical Jesuit gold. In 1990, Tumacácori became a national historical park with the addition of the mission ruins at Calabasas and Guevavi.

Before Karen and I left Tumacácori last September, we paused at the visitor center's 1937 replica of a mission-era garden and patio. I wanted to see the fig tree, a descendant of one of the original fig trees that Father Kino brought to the mission. We learned from a park official, however, that the tree had succumbed to a bad freeze a few years earlier. Although disappointed, I collected a few seeds from a cotton plant. Today, in a pot outside my home, shining white bolls erupt from tall, broad-leaved plants, just like those grown by the Jesuit and Franciscan padres and the O'odham people on the Santa Cruz River hundreds of years ago.

Tubac: Place of Rotting Enemies, 77 River Miles

I'm soaked with rain by the time we make Clark's Crossing about halfway between the mission and the Tubac presidio. Darkness seeps beneath the trees, and I feel both chilled and contented. The air smells rich with moldering leaf litter; shod hooves clack on stones and raindrops perform a rataplan on drumstick branches. No one speaks but the rattle-voiced ladder-backed woodpeckers.

Where the trail swings to the west side of the Santa Cruz, the horses step into the water without complaint this time. A gate leads us north again out from under the cottonwoods and into the mesquite, whose feathery leaves only partially conceal their dark thorny branches. The zigzagging branches of graythorn, at least, are more transparent with their warnings. The shrubs bear long wands of stout spines, each with a circlet of tiny oval leaves as if the greenery were only an afterthought. Graythorn is an important plant for birds, which build nests in the protective branches and eat the blueberry-like fruits.

The first indication that we've reached Tubac comes when the thick trees step aside to expose a broad open area. We keep to the right, following the river, quartering away from the field and the site of the Tubac Presidio that lies beyond it. My mind, however, leans in that direction and my eyes search for the presence of stacked adobe. I can't help but think that this open grassy plain might be the same place that Juan Bautista de Anza, newly promoted to lieutenant colonel, staged his great expedition that warm cloudless day in October 1775.

While Father Pedro Font recovered his strength at Tumacácori, Tubac thronged with activity. Adding to the 177-member expedition out of Horcasitas, another 63 people joined on. Some, like Sergeant Juan Pablo Grijalva, whom Anza had chosen to serve at the new presidio in San Francisco, had to travel with his wife and three children from Santa Cruz de Terrenate, the presidio on the San Pedro River more than fifty miles away. (In four years, unable to complete the garrison, the Spanish would abandon the site to the Apaches.) Others gathering at Tubac included muleteers, vaqueros, Indian interpreters, servants of the priests and Anza, and some thirty families (of both soldiers and colonists) with 114 children. Not only did the yearlong expedition require provisions for all these people, but the new settlement would need supplies as well. Pack mules, horses, burros, and cattle numbered nearly a thousand. The mules would carry six tons of flour, beans, cornmeal, sugar, and chocolate, a ton of cooking kettles, tools, munitions, and iron for horseshoes, and another ton of blankets, spare clothing, and tents—all of it packed and unpacked each day. This was the life support for a new and distant colony, much of it bled from the tenuous community of Tubac. Anza certainly already knew that before his return from California, Tubac would suffer the presidio's relocation to Tucson and that in this it was necessary for the tiny settlement on the Santa Cruz River to give its heart to the people destined for San Francisco.

I can see them in my mind on the day of their departure, a colorful horde heaving their trail into the sky, the smell of animal sweat and dung rising with the giant plume. Lieutenant Colonel Juan Bautista de Anza in his blue capote on horseback leads the way, followed by the saddle-slung, mule-riding priests, Pedro Font, Francisco Garcés, and Thomas Eixarch, the latter two traveling to the Colorado River to "ascertain the wishes of the tribes who live on its banks." Father Font sings hymns of praise in a high nasal tenor tremolo that the people walking behind him lift up and amplify in a harmony of excited voices. The songs drift beyond the mule train to the herd of horses and cattle with their vaqueros bringing up the rear of this mile-long itinerant town.

Anza had probably already arrived with his colonists at Monterey, California, when the presidio at Tubac transferred to Tucson in early 1776. Tucson claims its "birth" on August 20, 1775, because that's the day Colonel Hugo O'Conor, the red-headed, Irish-rebel inspector general of the Spanish army in northern Mexico, decided to move the garrison. Tubac had already perished more than once and would do so again, but it seems that this time Tubac bore from its dying body not one but *two* new communities.

I've descended the stairway into what remains of Anza's home at Tubac, a fortified house originally built in 1752 by Captain Juan Tomás Belderrain, the first commandant of Presidio de San Ignacio de Tubac. Standing in a narrow, darkened room, I can see the foundations of the "Captain's House," the park's oldest structure and the center of Arizona's first European community. Once, the Captain's House had thick, outer walls with gun-slit windows. And inside this wall, rooms surrounded a rectangular courtyard that functioned as home and military fort, holding trials, weddings, and fiestas. But these days, behind the underground exhibit's glass, all that remains are a few courses of gray adobe, the outline of an interior presidio wall, and the plaza floor. A tracing of dark ash tells stories of Apache raids or funerary pyres during times of epidemics. Bits of smooth shell hint of a trade network between the region's native peoples before the time of the Spanish, while cat's eye marbles and toys note the European children who passed through a later-constructed schoolhouse long after the Spanish were gone.

The first people known to farm along the Santa Cruz River here called this place "Tchoowaka." They were O'odham and lived in one of the many small scattered settlements that ran up and down the river. As was common, they constructed several compounds of a dozen or so houses that surrounded a central clearing. They raised these houses over pits dug into loamy sediments, setting wooden uprights and horizontal braces, ridgepoles and beams into a rectangular shape and then thatching the framework with grass.

Tchoowaka entered history on April 16, 1726, with the arrival of Father Joseph Agustín de Campos, a Jesuit missionary. The padre had stopped at the village "to take a siesta," according to the historian Henry F. Dobyns, and "he noted the place because he was cajoled into baptizing a nursing infant during his rest period."

Dobyns also has the most interesting of the many translations of "Tchoowaka," or "Tchuvaca," the Spanish rendition of the O'odham word from which we get the English "Tubac." He says the word translates into English as "rotten," and relates to an O'odham story about an enemy attack on the village in which several of the attackers were killed and their bodies left to decompose. The place name carries the connotation, he says, of "The Place Where Some Enemies Rotted." Probably won't find that on a Tubac travel brochure!

With the arrival of the Spanish and their diseases, with the growing strength of the Apache and European efforts to deal with them, the population of Tubac flexed and fluxed, sometimes severely, over the next centuries. In 1732, the village became a *visita* to the mission at Guevavi and apparently had a sizable Spanish population based on the number of unmarried Spaniards of marriageable ages. From 1745 to 1751, mission records for Tubac show eighteen baptisms, fourteen marriages, and only two burials. That only two recorded deaths occurred during these seven years when the rest of the region was suffering high mortalities from epidemics seems amazing; the death rate at Guevavi was about sixteen people per year.

In response to the Pima Revolt on November 21, 1751, Governor Ortiz Parrilla established a presidio at Tubac with a company of fifty soldiers under the command of Captain Juan Tomás Belderrain. The frontier town sprang to life. By 1766, Tubac boasted seventy buildings. Following Juan Bautista de Anza's California expedition and the relocation of the Tubac presidio to Tucson, however, the population began declining, the community reverting to "pueblo" status as people moved to Tucson or Yuma. One year, Apaches made off with Tubac's entire herd of horses and cattle, then returned a month later to strip its fields of corn. A smallpox epidemic in 1781 probably spurred Tubac's next abandonment, because two years later only empty clusters of decaying adobe marked Arizona's first European town.

Over the next fifty years, the population of Tubac would continue to rise and fall. Apaches would resume their raids whenever military forces waned, and settlers would again find it necessary to abandon the town. Tubac's population remained unstable as people passed through on their way to distant riches during California's gold rush, or as governments changed hands, this time Mexican for American. But Tubac would continue to be a place of firsts.

One story begins in 1854 with Charles Poston, a customs house employee in San Francisco, and a mining engineer named Herman Ehrenberg. After seeing the region's potential, the two formed the Sonora Exploring and Mining Company, which became hugely successful. In less than four years, Tubac became the center of culture for the territory. Charles Poston, the self-proclaimed father of the community, started calling himself the "Alcalde of Tubac," and began performing marriages, baptisms, and divorces, much to the chagrin of the Catholic church.

When William Wrightson arrived to help Poston with his mining company, he brought with him a printing press. On March 3, 1859, Arizona's first newspaper, the *Weekly Arizonian*, started publication in Tubac. The weekly paper, "devoted to the general interests of Arizona," covered news from Arizona, Mexico, and elsewhere in the world. The Tubac Presidio State Historic Park has reprinted copies of the first issue for anyone interested in reading about such things as the rumored sale of Sonora and Chihuahua by Mexico's president. My favorite is the news item about an "enterprising showman" on the East Coast and his exhibit to "crowded houses" of a company of trained fleas. "Their feats," the article says, "are . . . truly surprising."

One hundred years later, in 1958, Tubac added one more first: the old presidio became Arizona's first state park.

We turn away from the Santa Cruz River where it passes beneath a concrete bridge and lead our horses along a road toward Tubac. We soon find the horse trailer, parked next to a baseball field where Marge Izzo left it more than six hours ago. We're tired, sore, and wet but exhilarated. My legs remain stiff and bowed as I swing out of the saddle and drop to the muddy ground. Everyone dismounts in slow motion.

We assist Ron and Bill with loading the horses into the trailer, then tip the two wranglers and say our good-byes. The girls are already talking about the next ride, which, Ron insists, must happen in the Santa Rita Mountains. "Some of the most beautiful country anywhere," he says. We make promises to connect again soon and start walking into town to the historic state park where I left our car. All I'm thinking about is dinner at our favorite Greek-Mexican place, the Tumacacori Restaurant, across from the mission. You can't beat Maria's delicious Athenian chicken, and Robert is quick with the ice-cold cerveza.

Looking back on that short, ten-mile ride along the Santa Cruz River, I wilt considering the hardship Anza and his colonists must have overcome to survive a twelve-hundred-mile, yearlong journey to San Francisco, California. What propelled them to take such a stake on an uncertain future? Perhaps Juan Bautista de Anza was an even greater inspirer of men and women than he was explorer and soldier.

It seems fitting to me that the man would spend his last days as he had spent all his days: in the service of people. I think of Anza as a kind of secular Father Kino, although Anza was a very devout Catholic himself—he personally financed the construction of the first church at Tubac for his soldiers, the chapel of Santa Gertrudis, a reincarnation of which (Saint Ann's) stands beside the presidio site today and still serves its parishioners. Both men felt the calling of God on their lives and answered it in a far place that many probably believed God had forsaken. Where one built missions, the other built colonies. Both men were explorers at heart, opening new routes to new lands and new peoples. If Kino was the "Padre on Horseback," then Anza was "El Jefe on Horseback." If Kino was the hinge to the region, Anza was its door swung wide open.

After taking command of the Tucson presidio in the fall of 1788, his sudden and unexpected death at his home in Arizpe, Sonora, must have stunned the community. There's no question that people revered him. They have for generations and continue to do so. I've seen it at his burial place in Arizpe, at the Spanish mission Nuestra Señora de la Asunción. His bones hold a place of honor beneath the floor of the cathedral, held in a glass-topped coffin on the epistle side of the sanctuary. (Kino was interred on the gospel side of the chapel in Magdalena.) There, they are visible to all who come and care to look upon the simple remains of the man who gave his life for this land and its people.

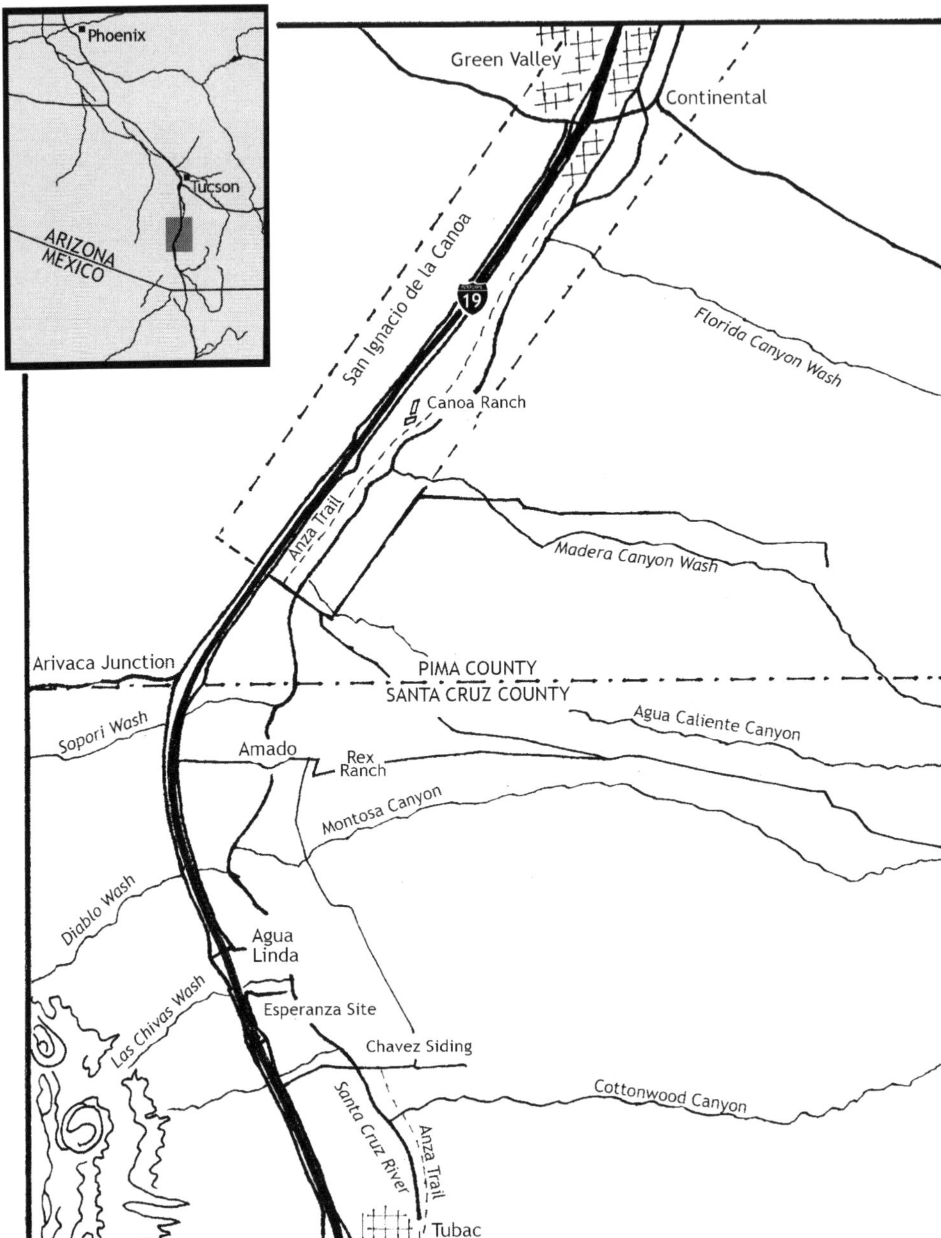

The Continental Reach

6

A River Underground

The Continental Reach

Hacienda de Otero: June, 79 River Miles

From the Anza Trail at Bridge Road just outside of Tubac, Karen and I hike downriver under tall cottonwoods and willows along the west bank of the Santa Cruz. It's two in the afternoon at the beginning of June and hot: 104 degrees *that's not a dry heat.* The river burbles past us, heading north toward Tucson, smelling of effluent and raising the humidity with protean clouds of gnats, which use our bodies as salt licks.

Water here means bugs, and I find myself looking forward to my latest objective: locate the place where the river slips beneath the sand.

In 1852, despite a wet summer monsoon season, U.S. Boundary Commissioner John Russell Bartlett noted that about nine miles north of Tubac, the Santa Cruz River went dry "notwithstanding all the rain that had fallen." River conditions haven't improved.

In recorded history, water has flowed the entire course of the Santa Cruz River only in heavy flood. Before the end of the nineteenth century, the river had surface water year-round in just five places—the most being in the San Rafael Valley and Sonora, Mexico, and here, near Tubac. Geology is what made the river perennial in these places, and in some places, like the San Rafael Valley, it continues to do so. Here, in the narrow valleys between mountains, shallow, impermeable bedrock and thin alluvium, like a monolithic birdbath filled with sand, hold water near the surface. Farther downstream, both Tucson Mountain bedrock and volcanic dykes, like those that form San Xavier's Martínez Hill and Tucson's Sentinel Peak, similarly once lifted the river's subsurface flow to create several marshes, or cienegas. Today, however, these cienegas are gone, their waters depleted as groundwater pumping and irrigated agriculture over the past sixty years have lowered water tables.

What I want to see now is what happens to the river when it begins to encounter the deep valley fill between here and Martínez Hill, hundreds of feet of gravel and sediments, washed in from the surrounding mountain ranges over many millennia. Will there be a giant sucking sound?

Karen and I walk side by side on a wide trail, finding relief in an occasional breeze. Male vermilion flycatchers and equally red summer tanagers zip around us as if someone were throwing ripe tomatoes, roma and beefsteak, respectively. The river remains on our right, pulsing in a synchrony of width and velocity as the landscape rhythmically squeezes and releases it. Water parsnip hugs the banks, lifting feathery leaves and umbrellas of tiny yellow flowers above our hips.

A tongue of cottonwoods, willows, and ash splits the second largest mesquite bosque on the Santa Cruz. The two habitat types—the broadleaf forest gallery with its understory immediately adjacent to the river and the bosque that borders it—are the rarest in the Southwest. This place is home to more than the flycatchers and tanagers. It's one of the few riparian areas in Arizona where rare birds like gray hawks and yellow-billed cuckoos come to nest. Tumacácori National Historical Park, three miles south of here, holds as many as nine cuckoo nests and a rare concentration of breeding indigo buntings.

Only in the past year along this reach of river, Karen and I spotted our first yellow-billed cuckoo, a shy relative of the roadrunner with a haunting *keow-keow-keow* woodland song. The same day also brought our first tropical kingbird, summering here from Mexico near a flooded cow pasture punctuated with white-faced ibises like spilt commas. But the real find was seeing an amazing roseate spoonbill, standing in a pond on one pink leg as beautiful and grotesque as a plastic lawn ornament. So today I'm expecting anything, maybe vulture-mimicking zone-tailed hawks, or thick-billed kingbirds snapping at insects, or tree-nesting, black-bellied whistling-ducks, or my recent favorite, the big-headed, rose-breasted becard, a Central American bird first discovered nesting on the Santa Cruz River in 1947.

Along with birds, the river corridor is home to many mammals and reptiles. Javelina and white-tailed deer browse on the fallen beans of mesquite, while mountain lions wait for a moment of inattention. As gray foxes hunt cotton rats and pocket mice, four kinds of skunks—striped, hooded, spotted, and hog-nosed—rummage through shrubs and leaf litter for rodents and insects, competing with raccoons and the tiny, kittenlike ringtail. The river's draw of insects also nourishes western pipistrelles, the smallest of bats, and a plethora of reptiles like whiptails and tree lizards, garter snakes and riverbed-probing mud turtles. The Gila topminnow, a fish endangered in southern Arizona, still ripples the quiet backwater pools of Tumacácori.

This wildlife diversity ranks this part of the Santa Cruz River with the San Pedro, Rio Grande, and Colorado rivers, as among the most sensitive and essential wildlife corridors in the Southwest.

The upper Santa Cruz River, although diminished, still retains the greatest value for wildlife and people. Fortunately, the potential is good for holding onto a healthy ecosystem here, mostly due to an unlikely source: effluent. Discharge from the Nogales International Wastewater Treatment Plant (NIWTP), spilled onto a complementary geology, gives life to this desert riparian forest. This direct flow, coupled with the natural runoff from surrounding mountains, recharges the aquifer and raises the water table to the degree that, despite the absence of any stone-striking prophets, springs emerge from the rocky ground.

Moreover, the river corridor is the least degraded. The river itself still snakes across the floodplain, not yet constricted by high banks or channelized as is so evident downstream near Tucson. Also, limited development has left significant open space, something local residents have recognized as important to wildlife and worthy of preserving. One of the goals of the Santa Cruz County Comprehensive Plan, adopted June 29, 2004, is the conservation of the Santa Cruz as a "living river" ecosystem. Part of this goal says, "The river, although damaged over time by human activity and grazing, has begun a slow recovery from its wounds. Land use decisions must allow this healing process to continue for the sake of the generations to come."

Karen lies on her back and closes her eyes, resting on trim, mesquite-shaded grass after eating a bagel with cream cheese and drinking half a bottle of water. We've lost the trail but found a golf course, this one with a nice restroom and chilled water from a cooler at the thirteenth hole. Once she spotted the facilities, Karen made a beeline for them, and now, seeing how peaceful she looks under the brim of her straw hat, I'm wondering if we'll make much more progress today. The trail slipped away from us after we crossed the river on a wooden plank and then encountered these widening pools of emerald green and the slim asphalt ribbons connecting them—the Tubac Golf Resort.

In 1789, a descendant of Spanish nobility named Don Toribio de Otero received one of the first Spanish land grants in Pimería Alta, amounting to about four hundred acres centered on these greens. Terms of the grant required Otero to build a home and live in it for four years, plant orchards and other crops, and support and serve in the military at Tubac, where his brother-in-law commanded the St. Rafael Pima Indian Company.

For the next seventy years, Toribio and his family fought with drought, Apaches, and squatters to scratch a living from the banks of the Santa Cruz River. But the withdrawal of soldiers from Fort Buchanan with the start of the Civil War in the early 1860s, leading to an increase in Apache attacks, forced the Oteros to retreat to Sonora. It would be only a few years, however, before this southern Arizona family of (currently) ten generations would once again begin building their legacy here. In 1867, Sabino Otero, Toribio's great-grandson, who became head of the family at nineteen, led the family's return to Tubac. With the advent of the railroads and his connections on both sides of the border, Sabino's ventures into livestock proved profitable and the Otero empire began to grow. Before he died in 1914, Sabino would become a cattle inspector, county judge, and a legislator. He would finance and build a school at Tubac, Arizona's first public school, and help found St. Mary's Hospital. He would earn the title "Cattle King of Tubac" with his seven ranches and seventy-five thousand head spread across the entire Santa Cruz Valley. His many descendants today say Sabino Canyon (El Canyon de Sabino) bears his name because of his ranch at the mouth of the canyon.

Teofilo, Sabino's younger brother, inherited the Otero estate after Sabino's death, but by then drought and competition had brought a decline in cattle ranching in southern Arizona. Teofilo eventually sold all his ranches except the original Spanish grant, where he moved to from Tucson. From the hacienda at Tubac, he finished his days as a gentleman of quality, one of the "Mexican elite," giving weeklong parties, traveling to Europe, and financing a school and church for the town he loved. After his death in 1941, and after 150 years of Otero presence, this last holding finally left the family.

Several people owned the fifteen-hundred-acre ranch afterward, including a colorful Nogales real estate baron and banker named Wirt Bowman, who was exceptionally successful at whatever venture he put his mind too. In politics he served as mayor, councilman, legislator, and democratic delegate; in business he owned First National Bank, built a club, racetrack, and casino across the border during Prohibition, and invested in Sunbru, the first bottled beer sold in Arizona after Prohibition. Then, in 1959, a group of businessmen, with Bing Crosby as their board chairman, acquired the Otero property and began construction of the Tubac Golf Resort.

I cross the greens to the original Otero hacienda, remodeled by Sabino Otero in the 1880s and now preserved as meeting rooms and a honeymoon suite. Huge multitrunked hackberry trees throw patterned shade onto red clay tiles and burnt adobe archways. Sculptured juniper trees frame the front entrance, whose doorway and windows hold lintels of dark ancient wood. A sprawling prickly pear cactus pushes its green pads across a corner of the roof, apparently finding nourishment in a thin layer of leaves decomposing between Spanish barrel tiles.

The first time Karen and I visited the place, a man pulled up in a golf cart and offered to give us a tour when I asked him if we might peek inside. His name was Dan Daly, and he was in charge of maintaining all the buildings. "We've tried to keep the furnishings and such to match the style of the time," Dan told us as we moved from room to room, noticing the leather-top boardroom table and accompanying chairs, the hand-carved wooden hutches, the punched-tin art. We walked over polished brick under high ceilings crossed with heavy beams. "This used to be all saguaro ribs," Dan said, pointing above us. "But it got termites. What I really love about this place are the doors and all the hardware on them. And look at this." Dan took us to the back porch to show us the roof tiles exposed from beneath, each one stained with a patina of blue between rough supports. "If you notice the tiles, they're all put together with leather. I've never seen anything like it. These days we use nails to hold them in place."

Just north of the hacienda, Otero's stables, hayloft, and bunkhouse have now become the Stables Bar and Grill, a gift shop, and business offices. Apple trees, possibly stock from the original orchards planted by Don Toribio under his grant terms, stretch their twisted limbs above a rusting corrugated metal roof of the barn, which no longer holds hay but exquisite pottery. Twin concrete silos, rising thirty feet to their clay-tiled rooftops, stand side by side like watchtowers transplanted from

Vernazza, Italy. Here, a Mexican restaurant named Dos Silos features delicious Baja-style fish tacos and a masa boat topped with beans, shredded beef, salsa, and goat cheese called Sope con Salpicon de Res. But the best part is the "real margaritas," which come with a warning: "These are VERY POTENT. Treat them with the same caution you would a martini." Every one of them always receives my sincere respect.

While Karen snoozes on the grass, I continue to wander about, enjoying the idyllic quiet. The fairways and greens provide fodder for a multitude of grazing, nibbling, snipping herbivores, all of which must be nemeses for the groundskeepers. Among the loose dirt mounds of gophers, I find piles of compact pellets dropped from deer and cottontails. By evening, droves of the rabbits will emerge from the shade to casually slip shoots of grass between their split upper lips. The spring-loaded furballs always make me smile, and not just because their frolicking, leg-kicking courtship dance looks like a vaudeville hat-and-cane act. Every year, the arrival of their young on our living room floor heralds the end of winter. And every time I remember "God's bunny."

Melissa's gray cat RainCloud is responsible for the arrival of spring in our home. He can't seem to resist the tiny, warm, brown-eyed Easter toys with the floppy ears and twitchy noses. Invariably, he carries them inside in the early morning to share with the rest of the family, usually dropping them on Karen's sleeping form. Sometimes they're even alive. On one occasion he delivered a healthy (uneaten) baby cottontail to our daughters, who decided that they should keep it and raise it, as it certainly still needed its mother, being so small. The girls' grandmother, however, who was babysitting at the time, thought otherwise, explaining to the girls that the rabbit didn't require their help. "This is God's bunny," she told them. "And God will take care of it." With a few more assurances, she persuaded Melissa, Kasondra, and Jessica to return the rabbit to the desert.

God, of course, can be as red in tooth and claw as nature. The next morning, the girls woke to find on the living room floor a headless baby cottontail. "Oh, no!" they cried. "RainCloud ate God's bunny!"

The rabbits hug the shade while ubiquitous mourning doves coo in the heat and Gila woodpeckers chatter and dive from tree to tree. A vermilion flycatcher, as brilliant as if cut from the pin flag at the thirteenth hole, chases insects among the spreading branches over my head. So *this* is Hole 13, I muse, recalling the Border Patrol agent's questions about how to find this place. "Whole what?" I asked, bewildered, when he suddenly appeared out of the undergrowth with his questions about a reported thirteen whole Mexicans. "You're not a golfer?" he said before disappearing again.

I think about this as I'm looking at two parallel tracks, two-by-fours, probably, buried into the grass and quartering away from us across the adjacent fairway. Is this the Anza Trail? Part of the resort's efforts to preserve the historic (Spanish and Mexican) integrity of the property? The irony rattles my teeth—the Anza Trail began in Mexico and still today Mexicans use it to "colonize" America.

Karen finds the Anza Trail marker, facedown on its post. "This must be the way," she says, leaning the sign against the restroom.

"How convenient to run the trail past here. They must have had you in mind."

"If they had me in mind," she says, "they would have included a massage table and young handsome Mexican with strong hands."

We head across the fairway to the sound of a train whistle, which coyotes echo with their own shrill cries. Joining the east bank of a river slowed by heavy green mats of watercress and duckweed, we leave behind the manicured Bermuda grass and Aleppo pines of the golf course to follow a trail along the river's brushy fringe. Willow and ash and cottonwoods return to screen the air of movement. Where Cottonwood Wash sweeps in from the southeast, thickets of giant cane hedge both banks, the highest branches of the exotic grasses bent under the weight of dozens of red-winged blackbirds whose liquid gurglings, checks, and zees add a pleasing dissonance to the chorus of water. I look back along the river, which shimmers like fish scales in the afternoon light, and I imagine that this is the way the Santa Cruz must have appeared when John Bartlett passed by here 154 years ago.

Among the river cobbles, Karen and I nearly walk over a dead animal I don't immediately recognize. What remains of its stiff charcoal and white fur has been pulled up over its head to its teeth-studded jaws as if the animal died trying to slip out of a tight turtleneck sweater. "Don't touch it," she says when I reach to turn it over. "Look at the tail," I say, noticing its naked, ratlike appearance. "Do you know what this is? I've never seen one before, not in Arizona anyway. A possum!"

"Poor thing," Karen says. "I wonder why it died. Maybe it got caught in a flood?"

"Look how it tried to crawl out of its skin. Interesting that we would find it near Cottonwood Wash."

I had recently learned from Scott Wilbur, who had just completed an Avian Habitat Conservation Plan for this part of the river, that the drainage from Cottonwood Canyon is a vital wildlife corridor between the Santa Cruz River and Santa Rita Mountains to the east, and that many animals depend on the drainage to move between isolated habitats. Wildlife corridors like Cottonwood Wash, connected to the Santa Cruz River, which in turn joins other corridors, can help moderate the problems caused by development-fragmented wilderness areas. The corridors allow species to improve their fitness by exchanging their genes with other populations. They also help species to disperse and recolonize formerly depopulated habitats or, as seems to be the case with the Mexican opossum, enter new ones.

The progenitors of the opossum Karen and I found may have begun life in their mothers' pouches somewhere along Sonora's Río de la Concepción. They or their offspring could have crossed into Arizona after journeying up to the river's headwaters at Sycamore Canyon in the Pajarita Wilderness. From there, Peck Canyon would lead the shaggy-furred, bulbous-nosed, prehensile-tailed marsupials through the Tumacácori Highlands and into the Santa Cruz Valley. It's not hard to imagine. Other Mexican species like Tarahumara and barking frogs, green rat and vine snakes, and Sonora chubs have entered the United States from Sycamore Canyon. Humans, too, from yesterday's Apaches to today's Mexican immigrants, have used the same route. The opossum is just among the most recent of aliens.

Before cameras captured two of the mammals on film in 1999, biologists at Saguaro National Park, which wildlife corridors like Davidson Canyon, Cienega Creek, and Pantano Wash connect to the Santa Ritas, didn't know that opossums lived there. Randy Babb, a biologist with the Arizona Game and Fish Department, originally thought that opossums, the only marsupials native to the United States, were exotics in Arizona. He believed they had escaped from people who kept them as pets, or that they were intentionally released. Randy calls one enthusiastic Tempe, Arizona, man from the late 1920s the "marsupial equivalent of Johnny Appleseed." "Raymond Hock in 1952 was the first to suppose that all Arizona opossums were introduced," he told me recently. "This was mainly due to the efforts of a Mr. White who imported opossums from Arkansas and released them in several localities statewide."

But after encountering the animals in northern Sonora in the 1980s, as well as in the mountains of southern Arizona, Randy began to suspect otherwise. He started researching historical records and gathering information from road-killed and camera-trapped specimens. "And many folks generously shared their observations. After I received enough records of opossums all across southern Arizona, I could see that they exhibited classic Mexican faunal distribution—overlaying nearly perfectly with green rat snakes and trogons." He now thinks that, along with javelina and coatis, the Mexican opossum likely arrived in southern Arizona only recently on its own.

At Chavez Siding, I'm thinking about another kind of "classic Mexican faunal distribution" after Karen and I turn back on the trail to find our car. Suddenly, we hear voices and snapping branches at the far side of the river. We stop and wait. Soon, a group of Mexicans heading north steps into the open. The leader is unaware of our presence for a moment but then he sees us. The immigrants freeze at his signal. Our eyes lock together. The moment stretches into several until I break the impasse by waving. "No la migra," I smile. Karen takes my hand, saying in a low voice, "Come on, Ken. Let's go."

When I look across the river again, the Mexicans—I guess only a whole nine or ten—are moving away in single file. The last man stops and turns toward me, raises his arm high in the air and smiles. I return the gesture. It's the one language we both clearly understand, one clearly fitting for Anza Trail travelers.

"Explosion of Ducks" Ranch: September, 81 River Miles

Three months later, Karen and I return to Chavez Siding to renew our search for where the river slips underground. The monsoon season, one of the best in years, has brought more than twelve inches of rain (all falling this past August) to Tubac, which averages sixteen inches in a year. Agua Linda Farm, two miles downstream from here, reported seven inches in one week. I'm thinking: *This is the way Bartlett must have experienced the Santa Cruz River that stormy summer in 1852.*

Brown water gushes across the road where I park, unwilling to drive farther even in our high-clearance Ford Escape. Karen figures I can walk across to pick up the Anza Trail, which should take me to Amado; she will hike south from there and meet me in two hours. At least that's the plan. It's 5:00 p.m.; it will be dark by 7:00.

On the east side of the river, I hop a barbed-wire fence and step into a pasture of wet grass, its riverside margin bermed against possible flooding. The berm provides a trail, and soon I'm walking on a wide path mowed through six-foot-high thickets of amaranth, flushing a coyote that lopes across the pasture while looking over its shoulder at this knapsacked trespasser. Where a wash enters the river from the west, three heavy-winged vultures grip the highest branch of a hackberry tree. A raw fleshy head watches me with one beaded eye as I pass beneath.

After I've hiked a half hour, the mowed trail disappears and the amaranth suddenly stands over me. I must part it with both hands and stomp a path. Wearing shorts is a mistake. I decide to leave behind the high ground of the cow pasture, weary of the painful weeds and occasional fences. I crawl through a low-branching mesquite, slip down a tangled bank, and step into the river. My legs burn from sweat-soaked scratches. But immediately across the river, I see a drainage I recognize: Las Chivas Wash.

Las Chivas Wash, which drains the northern canyons of the Tumacácori Mountains to the west, cuts a deep incision through La Esperanza Ranch, where a new three-hundred-acre conservation easement managed by the Tucson Audubon Society borders the Santa Cruz River for two miles. La Esperanza, now an eight-hundred-acre ranch, was originally part of the Reventon Ranch, or Reventón de los Patos (literally "the explosion of the ducks," referring to the sound the waterfowl make when taking flight), owned by Elias C. Brevoort in the mid-1800s. Brevoort, a onetime saloon keeper and Tucson's first postmaster, appointed in December 1856 and working out of Fort Buchanan, was also a "Southern sympathizer," one of many who lived in and around Tucson at the time. On January 8, 1862, Confederate president Jefferson Davis signed off the southern half of the federal territory of New Mexico as the Confederate Territory of Arizona. Southern control of the territory was short-lived, however, crumbling in April 1862 with the famous westernmost battle of the Civil War, the Battle of Picacho Pass. Colonel James H. Carleton, commander of the Union's California Column who fought at Picacho Pass, entered Tucson the following June to establish order. Employing two "confiscation acts" newly enacted by Congress, he promptly began seizing the property of several prosperous men deemed "rebels." One of them was Elias Brevoort.

When Arizona became a territory in 1863, President Lincoln needed to appoint a new postmaster for Tucson. Elias Brevoort had fled. His Reventon Ranch eventually became part of Sabino Otero's cattle empire, and it remained in the family's holdings until after Teofilo Otero's death in 1941. The property eventually came to Carlos Ronstadt, one of southern Arizona's most prominent cattlemen and another with a long family history in the region. Carlos built an adobe-brick home here, raised cotton, alfalfa, and cattle and named it Agua Linda for its "beautiful water." In 1957, the Ronstadts sold Agua Linda to Arthur Loew Jr. of Hollywood movie studio fame (that same year, film crews for the movie *Oklahoma!* used Agua Linda's cornfields as a background scene).

Today, the Loews raise grass-fed beef, sheep, and turkeys along with organic produce—tomatoes, chilies, potatoes, and squashes among other vegetables—at their sixty-three-acre, U-pick farm. Their "General Store" is a converted saloon, complete with plank floors and beamed ceiling. "It was once a watering place for the Loew family and their friends," explains Laurel Loew, daughter-in-law of Arthur. Among the obvious signs of merchandizing—freezers of beef, baskets of vegetables, dried gourds and herbs hung from rafters, and a produce scale—are remnants of the store's former purpose. A sign above the old wooden bar reads "Loew Tide." Another on the door says, "We'll refuse service to anyone still on a horse."

"People have been farming this valley for a long time," Laurel told Karen and me, after she joined us fresh from her fields with mud clinging to her long legs and sweat darkening her blonde hair. "But I'm not a farmer." Laurel's husband, Stewart, is responsible for what he calls the place's "reinvention as a destination farm." In the mid-1990s, Stewart returned to the original adobe hacienda Carlos Ronstadt built, abandoning a career in New York City to build a business while raising a family where he grew up. "Farming," he says, "is not a 'dot com' type endeavor with a quick in and out investment and profit. Farming represents a commitment to a long-term lifestyle."

In 1973, the Duval Corporation bought part of the Loews' Agua Linda property and named it La Esperanza. When Devon Energy, parent company of Duval, sold La Esperanza to real estate developer Jim Olson in 2004, the company stipulated that a conservation easement be created and managed by the Tucson Audubon Society.

Last February, Karen and I joined Tucson Audubon's Ann Phillips and Kendall Kroesen and a group of volunteers for a tour of the La

Agua Linda ranch

Esperanza site. Only a year earlier transactions had finalized between Audubon, Jim Olson, and Devon Energy, which would provide an endowment to Audubon for management and restoration work. That past fall, the work had begun.

Karen and I felt encouraged by what we saw at La Esperanza. We'd already witnessed the new housing developments crowding into the mesquite trees along the Santa Cruz around Tubac. We knew there would be new homes near here as well. Jim Olson has plans to build a low-density development on the remaining five hundred acres adjacent to the site. But with the three-hundred-acre easement in place, he's hoping to attract conservation-minded residents. It seemed to us that La Esperanza is an example of how developers and conservationists can work together toward building a better, more sensitive and sustainable community for both people and wildlife. An explosion of ducks may yet be a reality here.

Rex Ranch, 83 River Miles

While I slosh toward Karen in wet shoes and socks along the banks of the Santa Cruz, following in the footprints of immigrants, a gray hawk screams at me. And then another. Since I first heard them at the Anza Trail near Rio Rico, the rare riparian predators have become my signal species for my river ventures. Even where the cottonwood canopy has begun to thin, the birds haunt the occasional shadowed groves. Their pitched alarm calls give me more hope than warning, hope that the river will continue.

A western sun casts light among long lines of trees and the river hushes. Its water slows, meandering and turning back on itself or twisting into loose braids over a broad and shallow plain. The light is perfect, beautiful, laying itself like gossamer over every blade and leaf and dark, shimmering eddy. I pause to write a few words in my notebook but instead just stare and wait, pencil in hand. It's one of those moments I can hold by the stem but can't quite pluck.

Twenty minutes later I reach the river's end.

My map shows I'm just downstream from where Diablo Wash enters the Santa Cruz River from the west and Montosa Wash enters from the east, a confluence that joins mountain ranges like two arms spread wide with fingers opened and resting on the highest peaks. The cottonwoods have dispersed; groves have become single trees. Instead of grass-knotted sediments, I walk over smooth, fist-sized stones and coarse gravel, the river pushing itself along beside me in a slowly creeping tongue of foam. There's no giant sucking sound, only a spongy hissing.

Half a mile later, the sun has slipped far enough below the rim of the earth to leave the dry river in shadow. The air is as moist and cool as the flesh of a toad. Perfect hiking weather, though I'm beginning to feel a cramping in the muscles of my legs. At 7:00 p.m., Karen walks out of the darkness to meet me in the sand-dimpled channel. "About time," she says. "How can you even see anything?" Her massage idea has taken hold, and she has made arrangements for the two of us to spend two nights at nearby Rex Ranch. Tomorrow, while I'm hiking the Santa Cruz from Amado to Continental, she will suffer through a lazy brunch and hot-rock massage, followed by an afternoon soaking in the Cielo Health Spa, most likely in the company of a Terry Pratchett novel.

Sweet. I'm happy someone will be enjoying herself.

We leave the Santa Cruz River at the Amado Montosa Road crossing and drive east just over a mile to the ranch for dinner. "I'll order our food," she says, when I drop her at the missionlike entrance, "while you shower and change." I'm starved but this sounds like an order.

Our room is in Pima House, one of the first adobe buildings remodeled in the 1930s when Houston cattleman Rex Hamaker bought this part of the four-thousand-acre Montosa Ranch from a government trapper and rancher named W. Knibbie and began operating it as a dude ranch. The architecture is classic Josias Joesler, the Swiss architect who designed many residential and public buildings in Tucson, one of his most noteworthy being the St. Philips in the Hills Episcopal Church. He also created the original Ronstadt hacienda at Agua Linda. I noticed the same St. Philips Spanish colonial look in the mission entrance to Rex Ranch, complete with a mission bell. I don't know if Joesler is responsible for the entrance, or even my room, but I do know he designed the Hopi House I can see from my window. The construction is what many think of as traditional Southwest, a style for which we probably have Joesler to thank. During the late 1920s until his death in 1956, mostly working under the patronage of developer John W. Murphey, Joesler produced more than four hundred buildings in Tucson. He is responsible for popularizing styles such as the Spanish Colonial Revival and Pueblo Revival, but he also went beyond these to create uniquely eclectic styles drawn from his blending of historical traditions from Latin America,

Greece, and Italy. His designs and the materials he used—burnt adobe, aged wood, stained concrete, clay roof tiles—were intended to convey a sense of antiquity. He also incorporated features like open courtyards and Arizona rooms to accentuate the surrounding environment. The two-story Hopi House, for example, shows Joesler's favored rustic qualities. Mexican saltillo tile covers the floors. Exposed dark wooden beams span the ceilings to rest on adobe walls. An arched entryway opens into a solid stepped wall set with Mexican pots, the wall suggesting the stairway behind it that leads to an open rooftop terrace. Like many of Joesler's designs, this one typifies the architect's astounding ability to capture a sense of place.

Fresh from a shower, I walk past the resort's lighted pool, the first swimming pool built in Santa Cruz County, according to the literature. Everything is lush and green from our exceptional monsoon weather. Palm and Italian cypress drip with moisture. From under a bougainvillea, a Sonoran desert toad the size of a ripe mango squats in its skin and watches me with amber eyes. I find Karen seated inside the Cantina Romántica, the resort restaurant housed in the territorial-day, double-adobe structure. She's ordered me chipotle pasta, courtesy of award-winning chef Misko Gilliland, and a glass of Merlot. I will sleep well tonight.

La Canoa: Place of the Canoe, 89 River Miles

This morning, to our surprise, the Santa Cruz roars past where last night we walked this channel. Then, the river was a dry, rock-studded arroyo of sand. But now, due apparently to an overnight thunderstorm somewhere south of us, the river has become a corrugated gray torrent forty feet wide.

I remain on the east side of the river where Karen drops me and hike north through desert broom and amaranth. The channel of the Santa Cruz sweeps across an open, stony floodplain, punctuated occasionally with twenty-foot cottonwoods and Mexican elder. The cottonwood-willow forest I've been traveling in since Rio Rico, one of the rarest forests in the world, inhabiting less than 2 percent of the Southwest, has dwindled to scattered trees. Somewhere, a gray hawk calls, maybe the last one I'll encounter on the Santa Cruz River.

Directly west, the town of Amado sits beyond a line of mesquite rising against a flat, dark, overcast sky. Amado is named for pioneer Manuel H. Amado, a descendant of one of four brothers who left Spain in the 1790s to serve as presidio soldiers in Sonora and California. In the mid-1800s, Manuel Amado and his wife came to the area from Hermosillo and started a successful freighting business, which he soon expanded to include ranching, dairy farming, and operating a butcher shop. Life seemed to be going well for the Mexican immigrant and his family. But then in 1874, President Ulysses S. Grant set aside seventy-one thousand acres of reservation centered on the San Xavier Mission for exclusive use by the Indians. For the rest of the decade, the government allowed the decision to simmer and roil until early 1881 when the U.S. Indian Service began sending eviction notices to all non-Indians within the new reservation boundary. The recipients, most of whom were Hispanic, had lived on the land for decades, but the O'odham people had grown more and more resistant to these settlers, and Manuel Amado was one of them. The government confiscated his dairy farm and burned his family home.

After the eviction, Manuel moved his ranching business and his family farther upstream on the Santa Cruz to this present location and started again, calling the place "Amadoville." In 1920, his son Demetrio, who was the town's postmaster, changed the name to Amado.

I pass the Sopori Wash confluence on the right, the place where the eastern flanks of the Cerro Colorado Mountains meet the river on their way to the sea, at least during one of the better cloudbursts. Today, the wash is dry where the Santa Cruz flows past it, although the desert broom, bent like strung bows, indicate a recent violent discharge. More dune material for the Gran Desierto on the northern Gulf of California. Sopori Wash has forced the river channel to swing sharply to the east, cutting its bank nearly to the railroad tracks. A train with dozens of graffitied cars motors slowly southward, and I wave to the engineer, who returns my greeting.

Somewhere near here, I cross from Santa Cruz to Pima County, but it's only a line on a map and of no significance to the bounding horse lubber grasshoppers and jackrabbits I spook under circling turkey vultures tilted against the sky.

By the time I've hiked an hour, the river has drawn down from a gray flood dozens of feet wide to six thin fingers of foam grasping at sand. Ahead, the sandy river channel, which I suspect was entrenched sometime in the past, has become a straight, narrow ditch, lined with a few cottonwoods, seep willow, and yellow snakeweed, its banks held tightly with woven tangles of Bermuda grass. Flocks of common sulphur butterflies siphon moisture and salts from patches of mud, looking like tiny yellow sails on a chocolate sea. Humidity and sweat soak my khaki shirt, brightening letters reading, "Same shirt. Different day."

Beneath Elephant Head Road bridge, the clay pots of cliff swallow nests cling to the concrete supports as if slung there like spit wads. I smell something familiar: the rich organic aroma of my tomato garden, which I fertilize with a tea I make from dry guano—bat guano! In the mud at my feet, ridges of tiny oblong pellets rise directly below the bridge's expansion joints. I twist my head to peer into a seam and the darkness begins speaking to me in pitched whispers. I drop my pack and pull out my flashlight. Hundreds of tiny membraned faces look back. Pin-prick eyes and shar-pei noses, ears like seraphim wings. I stand in the Sistine Chapel surrounded by angels.

Mexican free-tailed bats are the most common bats in the Southwest. In the old mining town of Bisbee, Arizona, Mexican freetails leak from one-hundred-year-old buildings like the one above Café Roca, emerging from a hidden seam in pulses like escaping bubbles. In the ghost town of Ruby, Arizona, freetails plume out of a deep mineshaft like smoke. And, all over Tucson, the bats drop from roosts under bridges and stadiums to wheel across a crepuscular sky.

Mexican freetails usually migrate to Arizona in March, returning to Mexico in October, so these bats will be leaving soon, ending their all-summer, sunset-to-sunrise, 150-mile insect-feeding forays during which no moth, beetle, or mosquito is safe.

The hidden bats reinforce in me the idea that there's a kind of *seeing* that goes beyond what I do with my eyes, involving all of my senses, and more. This kind of seeing is more a participation, a penetration. To see the world in a grain of sand—to see its dimensions so clearly—that I become part of the picture. There are days when I don't get it right, when my sight is dimmed, when I allow my mind to form a patina over my view of the world and lose the wonder of the simplest things, like membrane-faced angels clinging under a bridge.

Some days, but not *this* day.

Leaving the bridge behind, I cross into one of the few legitimate Mexican land grants in southern Arizona: San Ignacio de la Canoa. U.S. Deputy Surveyor John L. Harris determined these present borders in 1880, which the U.S. Court of Private Land Claims confirmed in 1899. In 1900, another U.S. deputy surveyor, Philip Contzen, repeated Harris's work, replacing his wooden markers with stones chiseled "S I D L C" (San Ignacio de la Canoa) and placing them at the grant's four corners and every half mile along the perimeter.

Recently, Karen and I went searching for these stone markers, hoping that more than a hundred years hadn't completely erased all evidence of them. I parked about a mile east of this bridge near the intersection of Elephant Head and Canoa roads and, while Karen walked along Canoa Road, I hopped the barbed-wire fence to examine suspiciously large and upright rocks. Over the next hour, three people stopped to ask Karen if she needed help (she told me later she was tempted to accept a ride from the guy on the motorcycle). No one asked me if *I* needed help. When we returned to the car, I had about given up—I even rolled one stone over to check beneath it. Then, on a whim, I crossed to the east side of the road and there it was: "S I D L C SE COR" scratched into the black desert varnish of a rock not thirty yards from the car. We had found the southeast corner of the San Ignacio de la Canoa land grant.

"La Canoa," or "Canoa Ranch" as people now call this place, is named for its description, "El paraje de la canoa," the place of the canoe, a point on the Santa Cruz River mentioned in the journals of many early travelers, including Juan Bautista de Anza, Fathers Pedro Font and Francisco Garcés. The Anza expedition camped here on Monday, October 23, 1775, four hours after leaving Tubac, and this is the place where Manuela Piñuelas, the wife of soldier José Vicente Feliz, died in childbirth.

The next day, Anza wrote in his journal: "At three o'clock in the morning, it not having been possible by means of the medicines which had been applied in the previous hours, to remove the afterbirth from our mother, other various troubles befell her. As a result she was taken with paroxysms of death, and after the sacraments of penance and extreme unction had been administered to her, with the aid of the fathers who accompany us, she rendered up her spirit at a quarter to four."

Fathers Font and Garcés carried her body to the mission of San Xavier del Bac for burial. The boy, José Antonio Capistrano Feliz, would survive the expedition and grow up in the new San Francisco colony.

Southeast cornerstone of San Ignacio de la Canoa

Early travelers on the Santa Cruz River added La Canoa to their journals because the place provided water with "a little digging in the sand," as Anza notes. The "canoe" reference apparently comes from what historian Thomas Sheridan explains was a "hollowed-out log some enterprising pilgrim had placed there to water horses." John A. Spring, a native of Switzerland who came to the states to fight with the Union in the Civil War and in 1872 became Tucson's second schoolteacher, wrote in his diary that the namesake was a bit more practical, that a "Mexican settler . . . had built a large canoe, or flat-bottomed boat, upon which he crossed the river whenever the lower, or western, road to Tubac became flooded by the summer rains." Colonel Charley Sykes's fantasy steamboats notwithstanding, this is one of the earliest reports of actual navigation on the Santa Cruz.

The original La Canoa grant came to Ignacio and Tomás Ortiz, the sons of Agustín Ortiz who had settled at Arivaca in 1812. The Ortiz brothers initiated their claim for about seventeen thousand acres spanning both sides of the river between Tubac and the San Xavier Mission in 1820. But then came the land survey and appraisal for agricultural value, the thirty-day advertisement prior to auction, and the end of Spanish rule. About three months later (December 15, 1821), the brothers offered a successful bid of $250 for the now Mexican land grant of La Canoa.

La Canoa would remain their property even as the region moved from Mexican to American possession.

After the 1854 Gadsden Purchase, La Canoa became inundated with American squatters. Pete Kitchen lived here between 1855 and 1862 before moving south to Calabasas to raise his hilltop stronghold "El Potrero." Some constructed log cabins and farmed or ranched the bottomland. Others, like Missourian William Kirkland, milled lumber hauled down from the Santa Ritas for Tucson's market. Still other early entrepreneurs attempted to open stores, selling food and liquors to the "traveling public." But Apaches continued to thwart any permanent settlement.

In 1859, the Canoa Hotel opened for business, but two years later Apache raiders burned it to the ground, killing settlers, hotel guests, and Edwin Tarbox, the hotel's twenty-five-year-old manager. Raphael Pumpelly, a tall, golden-bearded geologist/explorer, who in a few years would become Harvard's first professor of mining, was traveling through southern Arizona at the time and wrote about stumbling onto the aftermath of what would later become known as the "Tarbox massacre": "The sides of the house were broken in and the court was filled with broken tables and doors, while fragments of crockery and iron-ware lay mixed in heaps with grain and the contents of mattresses. Through the open door of a small house . . . we saw a body, which proved to be the remains of young Tarbox. Like many of the settlers, the first Apaches he had seen were his murderers." Pumpelly and his companions buried the hotel manager and several other unidentified victims at the site, but no evidence of the graves or hotel remains today.

In 1876, just before his death at age eighty-five, Tomás Ortiz sold his grant to Frederick Maish and Thomas Driscoll. The two businessmen soon began stocking their new ranch with cattle brought up from Mexico and developing the original water source, accomplishing a bit more than "a little digging in the sand." In 1887, after the Apache wars ended, they formed the Canoa Canal Company, planning to construct a canal from the spring straight down through the Santa Cruz Valley to

Howell Manning Jr. Canoa Ranch House

the city of Tucson. Although the project never got that far, a mile of the thirty-foot-wide, cottonwood-lined canal was still in use in the 1920s.

Then, in 1912, Levi Howell Manning bought the 17,203-acre land grant. The onetime mayor of Tucson had come from Mississippi in 1884 at the age of twenty to work as a newspaper reporter and wound up making a fortune selling ice and electricity. Manning soon sold the northern half of the land grant to a rubber company and expanded the southern half to one hundred thousand acres. Due mainly to the efforts of his son, Howell, who took over management of the ranch in 1921, Canoa grew to become the largest showcase ranch in southern Arizona. Howell Manning strung fence to allow pasture rotation, dug irrigation canals for crops, constructed corrals and barns and concrete-lined pit silos to store feed, and laid out the longest feeding trough in the country (one-third of a mile), long enough to feed fifteen hundred cattle at once. Howell created a breeding program to improve the quality of their stock and raised prized Arabian horses with the help of two famous thoroughbred stallions, El Jafil and Saraband. In its heyday during the 1940s, Canoa Ranch hummed with the small-town comings and goings of welders and blacksmiths, mechanics and wranglers, merchants and teachers. As many as forty-five cowboys lived together in bunkhouses or in adobe homes with their families at the ranch headquarters. Manning cattle dotted the vast grasslands from the Baboquivari to the Santa Rita Mountains, over five hundred thousand acres of public and private land.

At Madera Canyon Wash, the muddy consequences of a rainstorm in the cloud-maned Santa Rita Mountains spill onto the dry sand of the Santa Cruz River. I'm surprised by the scene, the liquid mud, the quantity of it. I'm a long way from Madera Canyon, and the flow has come all that distance, pushing over six miles of hot rock and desiccated earth to be sucked up by the river channel. I haven't even heard thunder, just the rush of this unfathomable water.

"There are things one cannot know about canyons and floods," writes Craig Childs in *The Secret Knowledge of Water* after purposely trapping himself on a rocky ledge to witness the "infamous wave" of canyon floodwater during a thunderstorm. I've seen this infamous wave, many times. The last happened in the very canyon now emptying itself at my feet, as if Madera were a giant seine for catching rainwater and concentrating runnels and rivulets into a boulder-relocation program.

At the beginning of August a year ago, Karen and the girls and I checked into our favorite cabin in Madera Canyon, Kubo #4, for a few days of family retreat before the start of the fall semester. It was Karen's idea, a "regrouping" she arranged at least twice annually. This particular August we drove through a stark mosaic of scorched oaks and mesquite, bare rocks, and new green, the result of a terrific fire that had begun in nearby Florida Canyon and crawled into the mouth of Madera. Richard and Cora, longtime friends and residents of the canyon and owners of Madera Kubo cabins, had to evacuate their home, a beautiful construction of gigantic, hand-hewn timbers that Richard had harvested and hoisted into place himself. Fortunately, the fire never reached their place. Now, Richard was concerned about the rain.

"We've had eleven inches and none of it is coming down the canyon," he told us Sunday evening. "I told the forest service they need to get up there and check it out." Only a thin black ribbon of liquid soot stained the bouldered drainage.

At 2:00 a.m. that night, a thunderstorm rocked the canyon and I awoke to the deafening sound of white noise. Rain pelted the roof in waves while quick strobes of light turned the room green. I watched the dark ceiling, imagining the cabin washing away in foam and mud.

On Monday, I hiked the nature trail in the damp air and tracked birds. Painted redstarts flitted among the oaks, flashing brilliant red bellies and white wing patches and tails. Mexican jays called with harsh echoing voices. At Cora's feeders, lesser goldfinches and pine siskins clung upside down to her thistle bags, while rufous and broad-billed hummingbirds sipped sugar water between mad chases and jousts.

"Have you seen Mr. Flame yet?" Cora asked me in her wonderful Filipino accent when she found me standing in front of her gift shop waiting for the flame-colored tanager to make his appearance.

On Tuesday, rain fell all morning. I spent the time writing while Jessica, Kasondra, and Melissa poked around in the creek. In the early afternoon, the girls came inside and Karen began preparing lunch. Rain continued to fall. Suddenly, we felt the ground tremble as the canyon began to roar. I thought: *I hope that's not what I think it is!* Karen screamed and I looked out the window to see a huge wall of logs and boulders and ashen sludge crashing down the creek and coming right toward the cabin. "Everyone out the front door!" I yelled, carrying my laptop and running after Karen and our barefoot girls. I was sure the cabin was gone and we with it. We jumped off the front porch and didn't stop running until we reached Richard's and Cora's place across the road. Richard later said he heard the girls' screams and thought they were crazy for playing in the rain, but I remember the look of white fear on his bearded face as he rushed outside and collected my rain-soaked daughters.

When we returned to the cabin, the air smelled of wet ash. The flood had churned right past the back porch, stripping the creek bank of vegetation and pulling out giant rocks like loose teeth, leaving behind a thick gray coating that smeared the trunks of trees to a height above my head. Nothing looked as it had only moments before. Still ringing in my ears was the sound of it: "like gritting teeth and clenching fists." Just as Craig Childs describes. "It was the sound that angels make as their wings are torn off."

I turn west at Madera Wash, away from the fanning stream, to go looking for "La Canoa." Thanks to a new survey of the grant by Richard Willey, which I've marked on my map, I should be near the place Captain Anza first made camp after his expedition left Tubac in 1775. I'm not expecting to see much—certainly nothing above ground predating a century or so—maybe some rusting pipes and the half-buried remains of a concrete irrigation impoundment from the early 1900s.

I thread my way through tall, prickly amaranth and emerge speckled with sharp seeds into a floodplain scattered with mesquite and bleached with grass. The grass is thick, folded over to the ground in layer after layer by winds out of the south as if fingers had recently combed through and flattened every long stem. This, I decide, is grass minus the lips of cows.

I search among the mesquite and amaranth but don't locate anything I recognize as twentieth-century irrigation works. Walking north along the newly cut Anza Trail, which runs the length of the Canoa Ranch west of the river, I notice a line of mesquite to my left, rising from a berm layered in dead branches. Could it be the Maish-Driscoll canal? The ditch is huge, much larger than I imagined after a hundred and twenty years of erosion. The cottonwoods are long gone, but mesquite trees still tangle its grass-furred and sloping banks. I climb inside and follow the sky-hemmed channel, knowing from Richard Willey's notes that it will lead me to the Manning's Canoa Ranch headquarters.

Fifteen minutes later, I look up as the dark wooden shingles of an elongated roof appear through the mesquite. A pair of matching white stuccoed chimneys at either end of the gabled roof stand out against the weathered shakes, as does the brilliant stuccoed adobe of its walls. This is the 1935 home of Howell Manning Sr. And just beyond it is the shake-shingled, white-adobe home of his son.

We've just recently come to what may be the last chapter of Canoa Ranch's long history, which began unfolding following the sudden death of Howell's twenty-eight-year-old son. One night just before Christmas in 1951, a drunk truck driver weaving across lanes on the Nogales Highway slammed head-on into the truck driven by Howell Jr., instantly killing him and two ranch employees, Dave and Andrea Waldon. Howell

Sr. dealt with the tragedy by retreating to the bottom of the bottle and selling the land he loved.

Howell Sr. sold 128,000 acres of his ranch, along with all of his cattle, to Kemper Marley, owner of United Liquor Company, retaining only this southern portion of the original grant. Over the next ten years, he continued to sell off his land, while fighting court actions from his other son, Marklan Manning, to have him declared mentally incompetent due to his excessive drinking. By the time of Howell Sr.'s death on October 10, 1966, at the age of sixty-seven, Canoa Ranch had lost all its former glory as equipment rusted, the lake dried, and cottonwoods died. Sixteen months later, Howell's wife, Evelyn, would sell the last of the grant lands to the Duval Sierrita Mining Corporation, but by then nothing remained of the showcase spread except for this complex of abandoned and deteriorating buildings, hardly noticeable to the travelers speeding along the new I-19 freeway.

Afterward, the property changed ownership a few times, first going to the Pennzoil company, then to Phoenician Financial (part of Charles Keating's failed American Continental Corporation), then back to Pennzoil. In 1994, Fairfield Homes purchased Canoa Ranch from Pennzoil for $6.4 million. Then the builder announced plans to convert its desert grasslands and ranch houses into golf courses, shopping centers, and thirty-seven thousand new homes. The announcement stirred a loud response. The most vocal, a grassroots nonprofit coalition of environmentalists, astronomers, historians, archaeologists, and Native Americans among others, calling itself Amigos de Canoa, pledged to save the place for open space. Fairfield employees met head to head with the opposition in packed hearings. In January 1999, the Pima County Board of Supervisors voted four to one to deny Fairfield's rezoning request, the first major decision by the supervisors against a developer in twenty-five years.

There was a compromise, however. The county would allow Fairfield to build a golf course and five hundred high-density homes, whose bright red roof tiles I can see today, sprawled along the ridges and skyline just west of Canoa Ranch. Fortunately, about 80 percent of this place remained zoned for rural homestead.

Pima County now owns Canoa Ranch and plans to create a historical museum and park here. It will feature a conference/education center, an equestrian center, a campground and staging area for Anza Trail riders (which has already opened on this part of the trail, north of Elephant

Head Road), and a working ranch, whose bunkhouses will accommodate as many as sixty-four overnight visitors. The primary attraction will be the heritage area where the original adobe houses, ranch buildings, corrals, and pond will be restored to their 1951 appearance.

Recently Karen and I met with Kerry Baldwin, Natural Resources Division manager of Pima County Parks and Recreation, after we had spent the morning hiking across the ranchland for a "daylight hour foot reconnaissance," as my special permit called it. Kerry's wife, Tina, a tall woman with reddish-brown hair, introduced herself while Kerry unlocked the chain-link gate.

"The last time we were here, we found a family of bobcats over there." She pointed to an adobe building with a shining corrugated metal roof and the corral beyond it, a construction of stacked mesquite branches in excellent condition and one of the finest "retaque" corrals in Arizona.

Tina and Kerry led us to Howell Jr.'s residence, a single-story rambling structure of the classic early ranch-house style in southern Arizona. We stepped over red concrete floors through a kitchen hung with cabinets of Mexican cedar into a large room with beamed ceilings and a stacked flagstone fireplace. From the living room, we entered a narrow hall, the former entry to the house, which opened into a space I immediately recognized. "This looks like an Arizona room," I said. "See, I'll bet it was a porch later closed off with windows."

"Yes," Karen agreed. "And they just left the tree trunks on the inside."

This part of the house, the south wing, was built in 1935 for Howell Sr.'s two sons. The open porch fronted two bedrooms split by a single, shared bathroom. In 1948, after Howell Jr. married, he and his bride Deezie added the north wing with its kitchen and living/dining room connected by a breezeway to the original home. They built custom floor-to-ceiling casement windows, enclosing the porch and its mortared flagstone floor and wood-beam ceiling supported by posts of stripped pine.

Kerry said the county plans to restore Howell Jr.'s so-called Long House as a museum to interpret 1950s life of the Manning family. "The heirs, when they learned we were restoring the place, said they wanted to return the original furniture. We hope to give the place the period feel of the last of the big Anglo ranch houses."

Next Kerry and Tina took us to Howell Sr.'s "Big House," another single-story compound structure that may have incorporated parts of an even earlier dwelling. At about the time of Levi Manning's death in 1935, John W. Smith finished the design for Howell Sr. and his new second wife, Evelyn, and the two started construction on a home that, except for a few interior modifications, has remained the same for seventy-seven years.

We entered through a small courtyard tiled with eight-inch terracotta between sunken concrete planters and turned left into the kitchen. We roamed through the breakfast nook and adjacent interior bedrooms. From a T-shaped living room with its hardwood floors and marble fireplace, we headed for the master bedroom at the north end of the house, stepping down and passing through an enclosed porch paved with more saltillo tile and set with more concrete planters. Two large dead bougainvilleas raised thorny branches to the ceiling. (A photograph in a 1937 issue of *Architectural Forum* shows evergreens in these planters.) The master bedroom must have been Evelyn's pride and joy. The twin picture windows on either side of the fireplace would have cast morning light across the hardwood floors, the sunlight warming Evelyn's face as she looked upon the shimmering Santa Rita Mountains before stepping into her dressing room with its cedar closet and window-flanked vanity. Even the adjacent master bath had windows, which matched those of the dressing room—three rooms and nine windows. These were rooms with views.

"The home will interpret ranching traditions of the region after the Gadsden Purchase," Kerry explained. "And we'll refurbish the kitchen to serve special events and receptions." An exhibit in the dining room will show the intersection of Native Americans and homesteaders. Another in the living room will feature changes in irrigation and agriculture, including the construction of the Maish-Driscoll canal. One of the interior bedrooms will illustrate the railroad's impact on ranching; the other bedroom will show the evolution of cattle brands. Even in the master bedroom, there will be displays of ranch artifacts and equipment.

Before leaving, we visited the foreman's house, a traditional Sonoran adobe row house with a long continuous porch supported by pine trunks, and one of the oldest ranch structures, dating to before 1924. The county will use it to interpret early Spanish and Mexican building practices and devote its largest room to the Ortiz brothers and the history of the San Ignacio de la Canoa land grant.

As Kerry, Tina, and Karen made their way back to the gate, I stopped at a mud adobe wall whose lime-plastered sides had crumbled in places. On its top, I noticed some letters etched into the cement: "Mayo 14, 1946." Beneath the date was a name, which I thought must be a ranch hand or perhaps one of the ranch children who took a moment to immortalize himself in the fresh cement on the way to the schoolhouse sixty years ago. It read: "J. J. Vasquez."

The people working along with Pima County Parks to preserve Canoa Ranch understand that the place is a microcosm of our region's history and people. The goal here, as spelled out in the "Canoa Ranch Master Plan," is to bring to life the people who lived here, specifically the Mannings and the Mexican Americans who shared space alongside them. The plan is to keep the smell and feel of a real mid-twentieth-century Arizona ranch, down to its fresh horse manure, sweat, and dirt. "The telling of the Canoa Ranch story encapsulates many of the narratives of southern Arizona," it says. "When we understand the rich and diverse past and present of the river, the land, and the peoples, and the economy of this place, then maybe we can plan a future for Canoa Ranch that respects and honors its vital place in our community."

"It's still a few years in the making," Kerry told me that day. "But when we've completed the project, Canoa Ranch will be a destination spot for the I-19 corridor."

I hike northeast from the main ranch complex, skirting the tamarisk-bordered but dry five-acre lake built by Howell Manning in 1921. As part of the lake restoration, Pima County will build a self-contained wastewater package plant to supply treated water for the lake and to help reestablish the floodplain's riparian vegetation, a floodplain already populated with relocated burrowing owls.

When I reach the river channel, I see the dark green swatches of Keith Walden's southernmost pecan groves. Farther on, a gushing chocolate discharge from Florida Canyon Wash (also draining the Santa Rita Mountains) rushes between the giant trees into the river and tears away the high opposite bank under the Torres Blancas Golf Club. I smile at seeing white irrigation pipes like bent soda straws jutting from under the overhanging fairway. The river will always have her way. My smile soon dissolves, however, as I enter a hot swamp of mud, Bermuda grass, and head-high amaranth. After fighting the thickets for half an hour, I climb a steep berm to walk among the pecans—but again I'm thwarted. Beyond the crest of the berm, which has become a convenient place to stack dead pecan trees, a wide field of dry, prickly amaranth barricades me from the cool shade of the groves.

I decide the tangled bank of bulldozed naked trees will be easier to traverse, but I don't get far before I realize it could be just as easy to break a leg. I return to the river channel and its mud, beating back the amaranth with hands and feet. I'm red-skinned and blistered by the time I finally see the Continental Road bridge where Karen is waiting for me, six hours after dropping me near Amado. I'm happy to finally arrive at Continental, today's hiking destination, and to return with her to soak in the hot tub at Rex Ranch.

Continental and Green Valley, 94 River Miles

After an interruption of natural rubber from Asia during World War I, the Intercontinental Rubber Company planted the latex-yielding shrub guayule on 9,700 acres of the northern grant purchased from Levi Manning in 1916. The company drilled wells, built facilities for processing latex into rubber, and raised comfortable thick-walled adobe housing for workers (including a schoolhouse for their children), which would become the town of Continental.

After the war and subsequent disenchantment with guayule as rubber prices collapsed, the company shifted to growing cotton, mostly unsuccessfully. During World War II, German POWs worked the fields of what was called the Continental Farm. Then, in 1949, Queen Wilhelmina of the Netherlands, who owned a controlling interest in the company, sold it to Farmers Investment Company (FICO), established in 1946 by Keith Walden and Chicago investor Henry Crown. The two doubled the size of the cotton farm, rotating in barley and corn while experimenting with Spanish peanuts, wine grapes, and various fruit and nut trees and raising eighteen thousand head of cattle divided between two feedlots. From 1965 to 1969, after disease damaged the cotton fields, FICO planted four hundred thousand pecan trees on five thousand acres of river valley, creating the largest irrigated pecan orchard in the United States.

During high school in the mid-1970s, I worked summer jobs with Mexican immigrants in these pecan orchards, pruning trees and driving tractors to clear the ground of fallen branches and dead trees. My stepfather, Tom Brancheau, was FICO's comptroller, and he introduced me to Keith Walden and his son Richard, who were only too happy to send me into their hot, sweaty orchards for $2.45 an hour. I remember their cowboy hats and boots, and their western-accented stories. I also remember visiting the feedlots and how the stench from tens of thousands of penned cattle would rise in a brown cloud to drift on the wind and settle over a new community of white midwestern retirees, much to their chagrin.

The burgeoning, seasonally fluctuating population at Green Valley soon put an end to Keith Walden's cattle operation, or at least this is what my stepfather believed. Bovine methane and shirt-sleeved golfers don't mix well. As the so-called golden years of Arizona's cattle feedlots (mid-1960s and '70s) felt the squeeze of the Midwest market, FICO's cattle operation gave way to a different kind of pressure from the Midwest: country clubbers crowding the northern half of San Ignacio de la Canoa.

The irony is that Keith Walden, who died in 2002 at the age of eighty-eight, had only himself to blame, although I doubt he ever blamed anyone for anything. In 1961, Walden sold 2,900 acres of his land, the northwestern corner of the land grant, to Don Maxon, a Chicago developer. After Congress amended the National Housing Act in 1959 to provide federal funds for housing the elderly, Maxon began looking at land in states such as Hawaii, California, and Florida for potential sites. Because he had been caring for two aging aunts, and because he himself suffered from arthritis and the effects of childhood polio, he understood the health benefits of warm, dry climates. The Santa Cruz Valley in southern Arizona seemed ideal.

While Don Maxon worked out the finances with the FHA, his architect brother Norm searched for an architectural theme. Josiah Joesler's designs impressed him most, and after the Tucson architect suggested that the brothers visit the Spanish colonial town of Alamos, Sonora, Green Valley had its inspiration. The central plaza in front of the cathedral at Alamos, with its colonnade of white arches, white stucco, wrought iron, clay roof tiles, and brick streets would become a model for the new retirement community.

The first apartment fired up its air conditioner in 1964. By the end of the year, one thousand more were ready (today's East and West Villas), with laundries, parks, shuffleboard and tennis courts, six swimming pools, and a nine-hole golf course springing out of the bladed river bottom. A medical clinic, fire/police station, and restaurant would also serve the residents—all they had to do was come. When they didn't, Maxon soon realized that he couldn't meet the obligations of the mortgage, and by the summer of 1966, the FHA began foreclosure proceedings. A little more than a year later, Maxon sold his company's interest in Green Valley.

But the upstart age-restricted community, with some reorganization, survived. By 1972, it had begun to thrive, growing to more than three thousand residents, and in August of that year, Scottsdale developer David Williamson came on the scene, establishing Fairfield Green Valley, Inc. By the summer of 1977, so many thousands of retirees had complained loudly enough about the smell of FICO's feedlots that Keith Walden, whom many consider to be the father of Green Valley, agreed to close down his twenty-two-year-old operation. The cowman became the gosherd, the goose herder, replacing cattle with snowbirds.

These days around twenty thousand people live in Green Valley, with estimates of close to thirty-five thousand in the winter months.

Sahuarita, 101 River Miles

I return with Karen later in September to continue my trek along the Santa Cruz River by first attempting to climb El Saguarito, which on my map rises to the northwest of I-19 and Sahuarita Road in the midst of a housing development. El Saguarito is the center point of the north boundary of the original Canoa land grant, as redrawn by Richard Willey, and lies twelve miles (five leagues) from the road and historical approach to the San Xavier Mission. Like travelers from the past, we want to scramble to the top of the stony hill and see the mission towers as if for the first time on a long desert journey north.

Unfortunately, the development is gated, and we must satisfy ourselves with the view from the freeway overpass. All we can make out, however, is the brilliantly lighted marquee outside the Tohono O'odham Desert Diamond Casino.

We drive west back into the town of Sahuarita, where James Kilroy Brown owned a ranch in 1879. Supposedly, he named the place for its many giant saguaros lifting their arms skyward as if someone had shouted "This is a holdup!" and left the cacti standing there as a kind of weird joke. These are among the first saguaros you encounter coming north along the Santa Cruz, an indication that the river, after journeying half its entire length, has finally dropped out of the desert grasslands and into the Sonoran Desert.

In the early 1880s, Sahuarita Ranch was the hub of the local population, with its post office, a railroad station, and a stop for Pedro Aguirre's stage line, which ran between Tucson and Arivaca. But Apaches soon drove Brown and his family from his land and he sold the ranch. It wasn't until 1911 that the community recovered enough to need another post office.

These days, according to the 2006 population estimates, nine thousand people live in Sahuarita, a threefold increase since 2000. The town incorporated in 1994 and has since begun crowding the northern boundaries of unincorporated Green Valley. Where Sahuarita Ranch was once the focus of the community, Rancho Sahuarita has taken its place. The new three-thousand-acre master-planned development among the pecan groves on the banks of the Santa Cruz River offers "the perfect balance of solitude and activity" with its quaint cafes and Super Wal-Mart. The town expects to build 11,600 homes by 2015.

Karen and I walk along a paved walking/bike path between FICO's oblique lines of groves on the east and the roof tiles and brown stucco of Rancho Sahuarita on the west. The path skirts the river in two places where its deepening trench snakes outside of the pecans.

Where Karen stops to admire dozens of queen butterflies fluttering over a patch of wild sunflowers, I sneak through a barbed-wire fence, jump a dry cement irrigation ditch, and venture into the pecan grove. Deep, slick mud from a recent flooding holds the crescent stabs of deer hooves and cloverleaf impressions of coyote paws. To avoid collecting snowshoes of clay, I keep to the thin grass and hexagonal tiles of drying mud. Dime-sized toads pop from the ground with each few steps. Soon, the sandy river appears among the trees. Bermed and channelized, its course is the straightest and deepest I've seen it yet. I imagine water shooting though here in flood, exiting the groves as if from a fire hose, gouging the desert of the Tohono O'odham Nation all the way to Martínez Hill.

On the reservation's southern boundary at Pima Mine Road, I step carefully onto the ties of a railroad trestle and cross high over the river channel. Karen won't follow despite my urging her that it's safe. "What happens when a train comes?" she says, more a statement than a question.

For some time, I've been following this river into the Tucson Basin. But it is here that I begin to take hold of the Old Pueblo's broad valley, seeing that I am ringed by mountains: Facing the Santa Ritas to my south, I turn clockwise to the Sierritas, then the Tucsons, and the Catalinas on my north, then the Rincon Mountains, and finally back to the Santa Ritas. From Guevavi where I stood on the banks of the Santa Cruz with Melissa to this spot at Pima Mine Road, the river course has dropped almost eleven hundred feet in elevation, more than the Mississippi River drops from Minnesota to the Gulf. But what does the Mississippi know? It runs south, like most rivers.

I stand on the steel trestle and look at the dry river forty feet below me. Before 1940, a wagon road ran here, connecting Tucson to Nogales on a route parallel to the river, which flowed only seasonally as much as a mile east of here. The road, however, intercepted and collected runoff from a series of arroyos coming west out of the Sierrita Mountains. Eventually, the road eroded into a deep cut that migrated upstream to capture the main channel of the Santa Cruz, leading to this present rerouting of the river. It wouldn't be the first human-caused rerouting of the Santa Cruz. I look at its banks, ragged and steep, bands of river sediments crumbling into the sandy bottom. I see history in the terra-cotta layers, and I see hope.

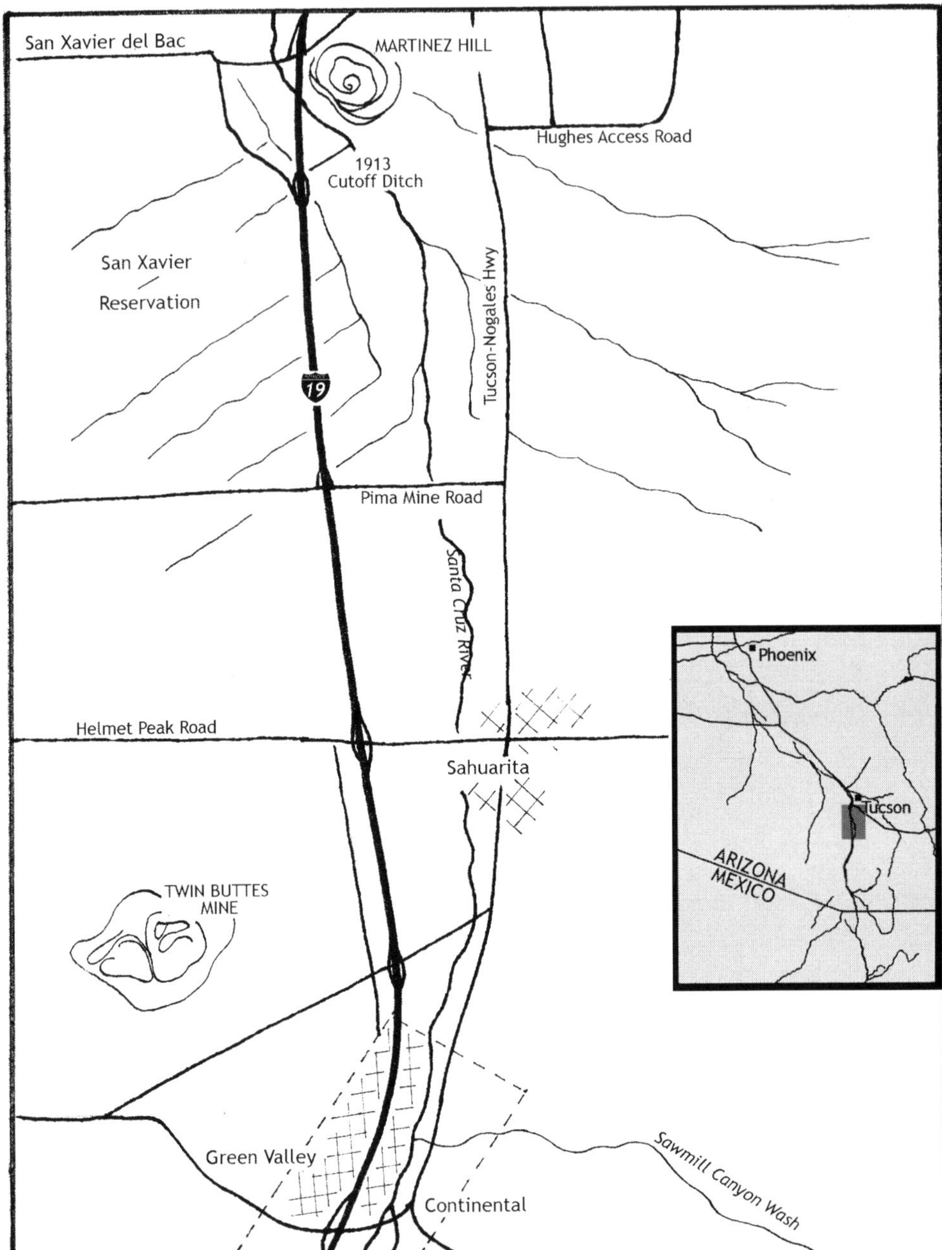

The San Xavier Reach

7

Native Water

The San Xavier Reach

Martínez Hill: June, 113 River Miles

Nine months later on a June morning of what meteorologists are saying will be a record-hot, 108-degree day, I return to Pima Mine Road and the river. Joining me is Danita Rios, a twenty-six-year-old Tohono O'odham woman assigned by the San Xavier District Council to "monitor" me.

Two weeks ago, I met with the six-member council to argue my case for hiking across the reservation. It wasn't an easy sell. The men and women grilled me on my purpose, questioning my intentions. "Why exactly do you want to hike through our lands?"

A month earlier I had written a letter to Austin Nunez, chairman of the San Xavier District of the Tohono O'odham Nation, asking for permission to hike along the Santa Cruz River across tribal lands to the mission. I expected a permit, some signed form like the one I received from Pima County to hike across Canoa Ranch. Instead, Austin Nunez wrote back saying I needed to make a request to the district council and that he had asked that I be placed on the agenda for the next meeting. "If they agree," he wrote, "they will require you to hire a monitor (guide) who must be with you at all times during your hike through our lands."

At Tuesday's meeting, I came up first on the agenda. Speaking in a microphone to the members, I answered their questions about my project and agreed to hire a tribe monitor, share a draft of my writing with the council, and send a copy of the book after its publication. The council voted 5 to 1 to grant my request.

With Danita close behind me, I slip through a barbed-wire fence and begin walking across the San Xavier District, home to about eighteen hundred people living on seventy-two thousand acres in the eastern portion of the Delaware-size, three-million-acre nation. On my right, the Santa Cruz River cuts a sheer-walled trench forty feet into the loamy desert soil. Nearly eight miles distant, the volcanic mound of Martínez Hill rises above mesquite trees and cholla cacti. Our destination lies four hours away.

Happy New Year, I think about wishing Danita when I notice fruit on the saguaro cacti splitting open and falling to the ground like discarded pairs of wax lips. The June ripening begins Hasan Bakmasad, the "saguaro moon," and the start of the O'odham New Year. It is the time of the cactus fruit harvest with its wine feast, a ceremony holy to the people. It involves songs and stories and dancing, along with the fermentation, drinking, and regurgitation of saguaro fruit syrup. The ceremony is a gift from I'itoi, the O'odham Elder Brother and Creator, who taught the people that the wine becomes a medium through which to pray for rain. The sacred intoxication and regurgitation represent "throwing up the clouds," and it is said that the saturation of the people with saguaro-fruit wine is like the desert becoming saturated with water, that during the driest and hottest and leanest time of the year, this is how the people summon the summer thunderstorms.

I'm certainly praying for an early Jukiabig Masad, or "rainy moon." At seven o'clock, it's already ninety degrees under a hard, azurite sky. The desert here, probably some of the most inhospitable on my journey, is baked ground, scored by decades of runoff erosion into deep, steep-sided gullies that we must continually backtrack around, moving farther and farther from the river. Creosote bushes look like spindly-legged tarantulas dying on their backs. Even their shadows wither in the sun. This is not the river valley of the past.

According to Texas A&M geosciences professor Michael R. Waters, prior to about eight thousand years ago, this part of the Santa Cruz River was a shallow, braided stream, broken into separate reaches and organic-rich marshes, or cienagas, which deposited sediments over a broad floodplain. Then, and for the next twenty-five hundred years, the river's character changed and it began eroding its banks and carrying away its sediments, significantly widening and downcutting the valley bottomland. Between 5,500 and 2,550 years ago, the Santa Cruz again went through a period of depositing sediments and saw the reemergence of its cienagas. Although there have been times when the river

Santa Cruz River at the O'odham Reservation

was stable, for the past twenty-five hundred years this sediment flex and flux pattern has grown in frequency, repeating itself with six periods of channel cutting and channel fill. With the drought of the 1890s (but possibly beginning as early as 1849), the river started another cycle of degradation, which continues to this day.

Waters, who in the 1980s studied the alluvial sequence on the river here, originally attributed the causes of these changes not only to climate but to impacts from human settlers, including those initiated by the canal-building Hohokam, ancestors of today's Tohono O'odham.

I admit I'm not a geoarchaeologist and do not fully comprehend all the nuances, but there appears to be a relationship between the river's erosion/deposition cycles and its human presence. Southwestern archaeologist Stephanie Whittlesey says that the Late Archaic period (between 3,500 and 1,650 years ago) "witnessed a veritable explosion of occupation in southern Arizona and associated changes in settlement, subsistence, and human organization." During this time, she says, communities grew in size and became relatively permanent, shifting to rivers due to "the increased importance of maize agriculture." Many of these river settlements were of considerable size. Several in the Tucson Basin, like the 2,200–2,400-year-old Santa Cruz Bend site, held two hundred dwellings grouped into smaller "villages" of houses arranged in a circle. Although little is known about water control in the Late Archaic, recent discoveries near Tucson show that farmers were digging wells to tap water tables and diverting perennial and floodwater into canals. These canals are the oldest known in North America, predating others by a thousand years.

The Dairy site, downstream from the Cañada del Oro confluence in northwest Tucson, shows early evidence of the larger, skillfully engineered canals characteristic of the Hohokam. What began as simple shallow ditches soon developed into miles and miles of adobe-plastered canals up to thirty feet wide and ten feet deep, the earth dug with stone and wooden tools and carted away in woven baskets.

And even right here, where I'm hiking today south of Martínez Hill, some scientists believe the Hohokam may have dug a headcut, a channel intended to intersect the water table and carry water into irrigation canals, that altered the course of the entire river. For some geohydrologists, this would explain why the river doesn't follow a more natural, geologically sensible course along the western margin of the basin.

With the advent of this culture around AD 700, an increasing population and reliance on irrigated agriculture must have placed substantial pressure on the Santa Cruz River. By their peak between AD 900 and 1300 (William H. Doelle, an expert on the Tucson Basin Hohokam, estimates the population rose to six or seven thousand around AD 1000), these ingenious people were irrigating thousands of acres of desert in the Santa Cruz Valley. Interestingly, Michael Waters wrote in 1988 that, following a time of stability, "a degrading river accompanied by erosion of the floodplain existed during the A.D. 1000–1150 period." This was the height of the Hohokam population expansion.

Waters today says some of his ideas have changed, and he blames the river's prehistoric downcutting on climatic causes only, not the Hohokam. But it intrigues me that there may be a relationship, a cause-effect that traces a history of the rise and fall of the Santa Cruz River in conjunction with the rise and fall of a human presence here, possibly beginning as long ago as twenty-five hundred years. Just the idea gives me hope, hope that the current "degradation" of this part of the Santa Cruz isn't permanent but part of a natural cycle, that the river will return again as it always has—once we move to more sustainable ways of living along it.

At seven thirty, we take our first break. Danita's black hair, which she has pulled into a tight bundle at the back of her head, is gleaming. I'm thinking that the silver stud poking through her lower lip must be heating up as well. But she is all smiles as we talk about her work for the tribe's cultural resource center, how she normally works with archaeologists on the reservation. "This is probably the first time you've had to hike for miles in one hundred degree heat with some crazy white guy," I tell her.

"Yes," she answers, "but I like seeing this part of the river."

The river has broadened, its sheer banks now several hundred yards apart, forming a terrace dotted with mesquite, blue paloverde, and acacia. A much narrower, sheer-banked channel carves a second terrace on our right, making a layer cake of the otherwise flat terrain. Looking at my map, I'm thinking that this must be the area where the Santa Cruz River once veered to the northwest, passing closer to the San Xavier Mission than it does today, into what we now call the West Branch. Somewhere ahead is the 1913 cutoff dike that carried storm water to the eastern Spring Branch and away from the mission's adobe walls and fields.

We head northeast to climb out of the river terraces and step onto a level plain studded with hundreds of mesquite stumps. Danita says that people once came here to cut firewood. Some of the trees, judging by the tabletop diameters of their sawn and weathered trunks, must have been gigantic. This, I think, must be the "noble woodland of mighty mesquite trees," described by the premier ornithologist Herbert Brandt when he visited here in the 1930s and '40s. In his typically florid style, Brandt writes of how the forest reminded him of a Sinaloa jungle, the trees like "grand old patriarchs" ruling over centuries of droughts and storms. By the 1960s, however, the ancient mesquite bosque had shrunken into stumps.

The cause of the demise of this mesquite bosque is complicated, but central to it was the rise of mining and agriculture in the valley, specifically groundwater overpumping from three main sources outside the tribe: the city of Tucson, the American Smelting and Refining Company (ASARCO), and Farmers Investment Company (FICO).

Today, the Tohono O'odham Nation is changing this. It began in the late 1970s when the tribe filed a lawsuit charging that Tucson and these companies had sucked water from beneath the San Xavier and Schuk Toak districts in violation of the 1908 Supreme Court "Winter's

Decision," which mandated a sufficient water supply for Indian reservations. The result came in 1982 with the Southern Arizona Water Rights Settlement Act. The act, however, required modifications before it could be fully implemented, and two decades of wrangling ensued, involving as many as thirty-five parties, including federal, state, municipal, and tribal governments, corporate entities, and private farmers. It seemed everyone in the state who drank or bathed or ran a hose to a garden plot jumped into the fray.

In December of 2004, President Bush signed the Arizona Water Settlements Act, allocating to Arizona's Native Americans nearly half (47.2 percent) of the 1.5 million acre-feet of water that flows through the Central Arizona Project (CAP) canal. Then, in 2006, Secretary of the Interior Gale Norton resolved the original complaint of the two Tohono O'odham districts by signing the Southern Arizona Indian Water Rights Settlement. The agreement meant that Tucson would pay to repair sinkhole damage and that both ASARCO and FICO would reduce groundwater withdrawals affecting the reservation. The districts would receive at no cost an annual 37,800 acre-feet of water from the CAP and 28,200 acre-feet from other sources, along with a $15 million trust fund to develop their water resources. The Nation's cooperative farm, once dependent on the seasonal vagaries of the Santa Cruz River, would now sprout green with water drawn from the Colorado River more than two hundred miles away.

Danita and I follow a cow path that drops off the former mesquite plain into a riverbed pinched tightly between fifty-foot-high walls of earth. Mesquite trees grip the loose rim for purchase, their twisted and shredded roots clawing at the crumbling banks or flailing in the open air. In places we walk in shade.

Soon we spook a pair of owls, an adult and juvenile great horned, which hopscotch ahead of us along the right wall, slipping soundlessly in and out of occasional dark crevices. "I hope this isn't a problem for us," I say, tentatively, recalling some Native American mythology about the night birds carrying off the souls of the dead.

"Some people say it could be. Owls are a bad sign. But," she says, offering encouragement, "maybe it's because we're having two funerals today."

At eight thirty we rest again, finding shade in the river's broken chocolate walls just beyond where an equally large drainage enters from the southeast. I lean against a giant clod of dirt the size of an adobe house and look at my map. It tells me this is where the entire eastern half of this valley discharges into the Santa Cruz, what was once called the Spring Branch. I'm thinking that the wagon road's accidental rerouting of the main channel into this branch must have made the 1913 cutoff ditch obsolete. For after 1940, the Santa Cruz River would no longer flow toward the San Xavier Mission (and into the protective cutoff dike) but along a course closer to the dark shoulder of Martínez Hill where it flows today, when it does flow.

Martínez Hill, our destination that I've kept in my sights all morning, is named for José María Martínez, who filed for a land grant between the river and the volcanic outcrop in 1851 when this place was still Mexico. His heirs were the only non-Indians allowed to remain on their land after the formation of the reservation and subsequent evictions in 1881. Apparently, José and his children were good neighbors. The grave of José's son, Nicolás Martínez, lies in front of the mortuary chapel in the original (1797) O'odham cemetery at the southwest corner of the San Xavier Mission church. His gray marble obelisk, inscribed "NACIO EL 5 DE ENERO 1848 MURIO EL 19 DE MAYO 1885," is the only headstone remaining in what is now a cactus garden.

At 10:00 a.m., our faces red and salted from a hot slog on loose, shining sand, we climb over a fence and enter a cottonwood grove where the Santa Cruz bends west as it bumps into Martínez Hill. I smell water, green and wet, and then see a dark ribbon soaking the ground. We drop our packs in shade, reveling in a drop in temperature of maybe fifteen degrees.

While Danita rehydrates, I pull out my camera to record what my eyes can't believe I'm seeing. Trees—tall, cool, shimmering cottonwoods and ash, bordering a slow-moving slip of water. I'd heard about the O'odham project, but this is unexpected.

Next to a ramada of ocotillo branches, a large sign reads: "Riparian Restoration Project." It says that the project has "re-created a native wooded plant community and wetlands" and that the area represents a river terrace of the former Santa Cruz River. "As the river is now dry," it goes on, "there is no longer enough water to keep the plants alive. The plants you see here must be irrigated with Central Arizona Project water. This water comes all the way from the Colorado River."

O'odham restoration site

Earlier, I had spoken with Mark Briggs, a restoration ecologist with the World Wildlife Fund, who helped design the site in 2001 and then implement it in 2003, describing it as two wetland ponds with a ring of cottonwoods, willows, and mesquite. "It is only a beginning," he said, explaining that the tribe has begun another restoration site farther upstream. "The San Xavier District has plans to restore the river here," he told me. "The elders remember what it was like not too long ago."

Even more substantial is the use of CAP water to restore the district's farmland. What were once a few small plots of corn and squash on either side of I-19 is becoming a thousand acres of cotton and wheat, alfalfa and heirloom vegetables—all free of pesticides and artificial fertilizers—under a $23.8 million farm rehabilitation project. The San Xavier Cooperative Farm will benefit from new irrigation systems, a high-efficiency infrastructure of piped water, improved roads, and flood protection.

I like to think of the Tohono O'odham Nation as a model for sustainable living on the Santa Cruz. One of the tribe's main priorities is to develop a water management plan that includes a large-scale groundwater recharge operation within the district. You won't find water sprinklers soaking golf courses here. You'll be challenged to find even one or two small Bermuda lawns for that matter. These people understand better than most of us that we live in a desert and that water is a limited resource. They have understood this for thousands of years, since the days they farmed along the river terraces at the foot of Sentinel Peak (A Mountain) where Tucson had its beginning.

Admittedly, the Tohono O'odham model is quite small, especially when we're talking about a human presence of a million people in the Tucson Basin alone. Small but not insignificant. We can do better by simply raising our consciousness concerning how we treat water, as we did before the CAP with Tucson's "Beat the Peak" program, which aimed at lowering peak demand. It's an extreme example, but I know two Tucsonans, brothers Rodd and Brad Lancaster, who live "off-main" in a comfortable home—with a garden yet—supplied entirely by harvested rainwater. This, in a desert that receives less than twelve inches of precipitation each year.

The White Dove, 113 River Miles

Scrambling over boulders to get out of the river channel at last, Danita and I walk northwest toward the beaming "White Dove of the Desert," the Kino mission of San Xavier del Bac, which the Jesuit founded in the native village of Wa:k, "where the water appears."

At the mission, Danita's work is done. I thank her and remind her to drink lots of water for the rest of the day. She says she's going to bed, and as we part I'm thinking she will be happy to get rid of this crazy gringo who hikes in the desert heat.

Outside San Xavier's ornate façade and rough wooden doors, I look up at Bartlett's "queer looking creatures."

Two hundred years after Father Kino established this mission in 1692, and sixty years after the tall, gray-eyed Basque, Father Juan Bautista Belderrain, began construction of the church in 1783, John Bartlett described the mission as "the largest and most beautiful church in the state of Sonora."

"It is elaborately ornamented inside and out, and contains many decorations new in architecture, partaking neither of the Greek, Roman, nor Gothic orders. Along the eaves is a row of queer looking creatures, the like of which cannot probably be found, even in the country of strange animals."

I can't believe that, as an artist and romantic, John Bartlett didn't immediately recognize the powerful symbolism employed by the Catholic mission builders. This symbolism carries so much more than words between peoples of two worlds whose translations may have lacked the deep and profound implications of faith. Pictures worth a thousand words in any language. No one could mistake what faith meant to St. Fidelis, the Capuchin martyr, whose statue high in the east transept shows a bleeding head wound and a knife jutting from his chest.

Perhaps Bartlett was referring to the two lions of Castile, standing upright against branches of lilies high on the façade. Or the headless figure of St. Cecilia, patron of musicians, whom tradition says lived for three days after an attempted beheading; or the statue of St. Agnes of Rome, patron of chastity who became a martyr at twelve or thirteen. Or maybe he meant the curious, weather-smoothed cat and mouse, the pair facing each other from their perches inside the curls of scroll at the far edges of the third tier. I wonder if he knew about the legend, that if the cat should ever catch the mouse the world would end.

There's no doubt, seeing the large sculpted shell above the balcony window of the façade, directly over the doorway, that I am entering a Spanish mission. The seashell is the symbol of the patron saint of Spain, St. James the Greater. The faithful make pilgrimages to his shrine at

Compostela on the coast of Spain and return with shells as tokens of their visit. Like other Kino missions, this church is full of shells. Shells hold basins of holy water and pedestals of statuary. They fill the domes of niches and sanctuary windows. One shell lifts the altar above the floor, while another, high above it, spreads its fluted valve like rays of the sun over the image of God.

As I enter San Xavier Mission, I pass through cool air smelling of wax. Sunlight dims and changes color where it strikes the high stained glass, the atmosphere attenuating to shining points where burning votive candles send vapors heavenward as prayers of the people. My eyes follow a painted cord around the right side of the sanctuary to the east transept, noticing that the same cord winds around on my left into the west transept. Like the belt of the Franciscans that tied their robes together, the interior of the mission is held together with a golden rope of knots. Hung with red draperies like dripping blood, the cord extends into the retablo mayor of the sacristy to fall on either side of St. Francis Xavier, for whom Father Kino had named his mission.

Wearing a white surplice over a black cassock, the bearded saint stands directly behind the altar in a place many Christian churches reserve for an image of the crucified Christ (who's there before Xavier's knees). At the highest point of the altarpiece, a red-headed, red-bearded God the Father lifts two fingers of his right hand while resting his left on a globe—an Irish God wishing peace upon the world. Immediately below Him is the Immaculate Conception with its regal image of the Virgin Mother, who stands in an oval cleft parted with red, undulating folds, suggesting an Alfred J. Quiroz painting of the Virgin of Guadalupe. Mircea Eliade had nothing on the sacred and the profane.

Father Juan Bautista Belderrain, after spending a couple months at Tumacácori, came to San Xavier in 1777 just before his thirtieth birthday. Here, he studied the native O'odham language under Father Garcés.

But by the time Father Belderrain started construction of the mission at Bac in 1783, his gray eyes must have lost some of their youthful luster. His friend and mentor Father Garcés was dead, killed by Yuma Indians two summers earlier. His brother, Ensign Juan Felipe, had tanked his career through a gambling habit, going from second in command at the Tucson presidio to one of its guardhouse inmates. (Juan Felipe would later redeem himself by reenlisting and paying his debts.) On one Sunday, Belderrain nearly went the way of his father, Juan Tomás, Tubac presidio's first commander who was killed by Seri Indians twenty-four years earlier. It seems the padre arrived at San Xavier at a time when open warfare had broken out between the Spanish and the Apaches, warfare that flared on May 1, 1782. That Sunday, coming up from San Xavier to say Mass at the Tucson church, Belderrain stepped into the midst of an orchestrated Apache attack that came close to wiping out both the presidio and mission.

This was the environment in which Father Belderrain, the religious right arm of an otherwise military family in Pimería Alta, would build one of the most magnificent Spanish churches on the new frontier. He wouldn't see it finished, however. Seven years after he laid the first fired-adobe bricks, he would die suddenly and alone in a pool of vomited blood, probably from typhoid. Supposedly, like his father at the Guevavi mission, he rests under the floor of the church.

Belderrain's replacement, Father Juan Baptista Llorens, labored the next seven years to complete the present church, mortuary chapel, and cemetery in 1797. Since then, for more than two hundred years, San Xavier, the Sistine Chapel of the Southwest, has served the Tohono O'odham people through the ravages of nature and nations as a refuge of history and worship.

In 1995, I visited San Xavier with Richard Shelton. It was his idea to stop by the mission during a bike ride that, unknowingly to me, would become a fifty-mile loop. We rode along Mission Road from our respective homes on the west side of Tucson, heading south toward the O'odham Nation. While Dick pedaled his red titanium-frame Cannondale touring bike, I tried to keep up on my hundred-dollar mountain bike. We left most of the traffic behind once we crossed onto the reservation.

Parking our bikes in front of the mission, we walked through its doors into a cool sanctuary whose six-foot-thick walls created its own weather. No one made a sound. I felt my sweat might somehow be an offense. What I remember most about that visit was that I'd forgotten how small the church seemed, but also how whimsical and colorful it was with all its Mexican baroque sculptures and pilasters, frescoes and murals, some gilded and others painted in sapphire and vermilion, cadmium and smalt.

Dick took me to St. Francis Xavier in the west transept, whose natural-looking statue (modeled after his 450-year-old "incorrupt" body in its glass casket at Goa, India) reclined in the midst of more than a hundred glowing votive candles. People in need of miracles had left rosaries and ribbons with names on them, small photographs of children or hospital bracelets—many pinned to the statue's coverlet. His forehead and feet had been polished smooth by the calloused fingers of thousands of supplicants, and I could not resist the pull of my own hand toward his peacefully sleeping face.

Today, only a few milagros grace San Xavier's light blue coverlet. But I still feel the familiar tug on my hand.

Immediately to the east of San Xavier I climb the Hill of the Cross to get a look at the River of the Cross as it continues north toward Tucson. On my left stands the magnificent White Dove, which, with its unfinished bell tower, has become a symbol of community for the O'odham and Hispanic Catholics here, as well as many others in the region who identify themselves with our rich and varied southwestern culture.

From the hilltop, I see new unplanted farmland stretching before me to Valencia Road, its wild geometry of furrows reminding me of 1880s hand-drawn maps of the fields that once filled the valley between here and Martínez Hill. Although today human technology wrests water from a distant river to irrigate this soil, I want to believe that these two volcanic outcrops might one day again raise the Santa Cruz to the surface to nourish the hard, immovable land of Bac.

It was the path of water that brought the Native inhabitants to the place, water for life and water for direction in life. It was the same for all who would follow them, whether Spanish or Mexican or American. Water creates community. And sometimes more. Writer Peggy Schumaker says that water in the desert is always holy. If this is true, then San Xavier del Bac, the place where a river rose out of a rock, must be as sacred as Horeb.

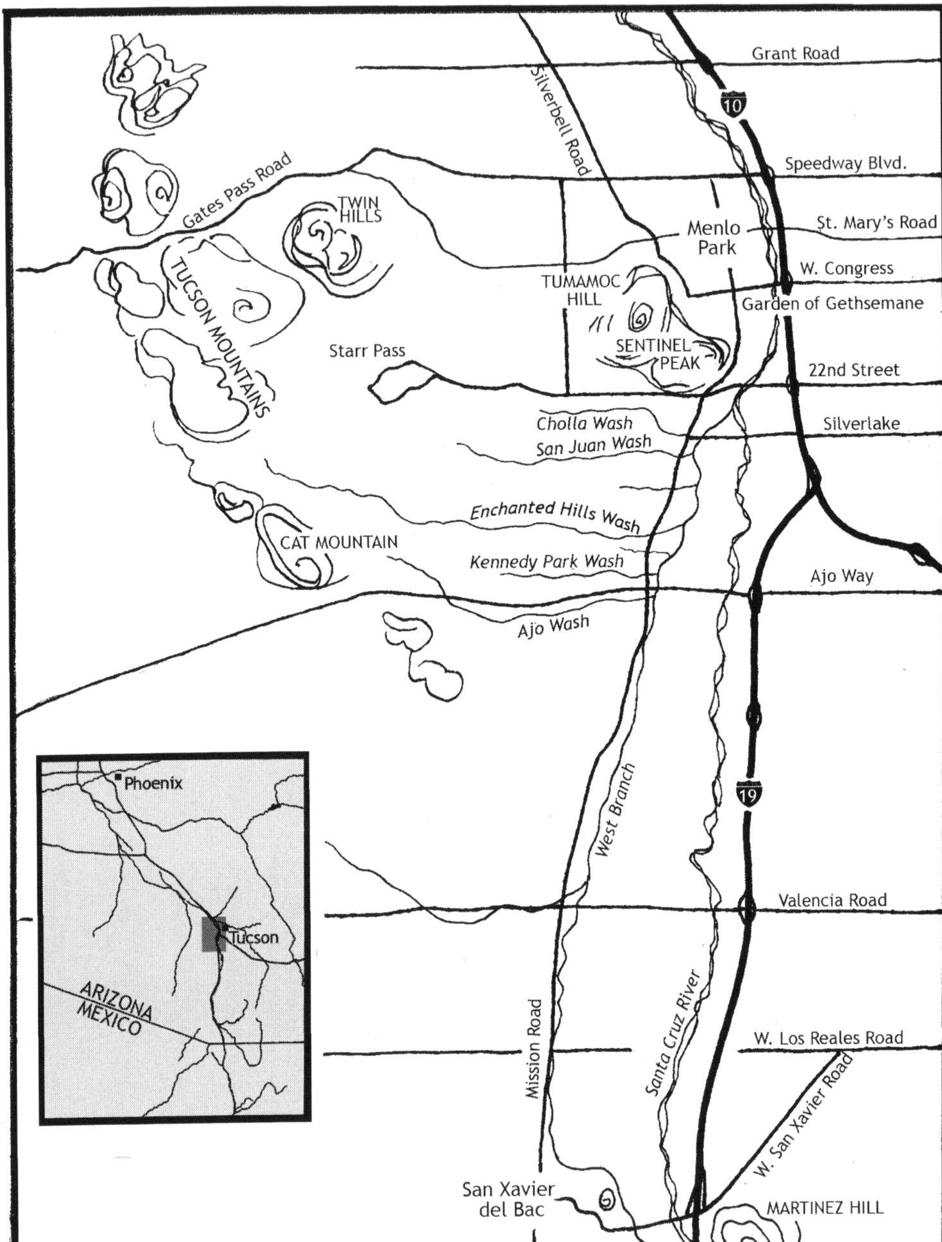

The Tucson Reach (West Branch)

8

New Rivers and Watering Holes

The Tucson Reach

Rocks have the oldest knowledge on Earth, rivers have the oldest names and determine where cities shall be built and what shall be their shapes.

—Alison Hawthorne Deming

The West Branch: March, 114 River Miles

On a cool spring day in early March, I return to the Santa Cruz River where it enters the fields of the Tohono O'odham, the rich brown furrows now green with wheat, courtesy of the Colorado River. Leaving the San Xavier Mission behind, I set Sentinel Peak in my sights, and decide to follow the original course of the Santa Cruz, what is now called the "West Branch." Before the 1880s, the river's main stem flowed here, closer to the church. The river's path to the east that we see today from I-10 was once called the "Spring Branch" because this is where the other of the two branches of the river meandered through a broad plain of cienegas and springs.

The Spring Branch, curling around Martínez Hill, became today's main stem of the Santa Cruz River beginning in 1888 largely because of a Welsh immigrant and developer named Sam Hughes. Hughes, who owned farmland along the river, had the idea that a twenty-foot-wide artificial headcut perpendicular to the river channel at St. Mary's Road would provide enough water to irrigate fifteen thousand acres. The headcut, essentially a large ditch, would intersect the aquifer and supply water even when the river itself was dry. But the river would soon undo his great plans. The following October, 1889, Pacific frontal storms dropped several inches of rain in as many days and the Santa Cruz began to rise, spreading mile-wide floodwaters over the plain as it always had. This time, however, the thin, sheeting flow encountered Hughes's ditch. An 1889 photograph by a German immigrant named Henry Buehman taken at this site, labeled "Falls on the Santa Cruz River," shows the result. With a backdrop of giant cottonwoods beneath Sentinel Peak, the picture looks like a muddy Niagara Falls as water cascades over wide crescents of eroding banks. The flood tore a deep gouge probably ten times the size and twice as deep as Sam Hughes's headcut.

Then, in the summer of 1890, an unusually cool equatorial Pacific Ocean steered tropical moisture into southern Arizona. Days of soaking rains followed. A series of floods in August started the arroyo clawing its way upstream, so that by the end of the month the waters had opened up a trench two miles long. The *Arizona Daily Star* on August 9, 1890, predicted the loss of hundreds of acres of farmland, saying "there are now several channels being cut by the flood . . . these new channels or washes are spreading out over the valley." Two years after Hughes dug his ditch, the Santa Cruz River had become the "Tucson Arroyo," a deep gash through the heart of the town from St. Mary's Road to a beautiful recreational impoundment known as Silver Lake (where Silverlake Road crosses the river today). It would take only a couple of decades for the arroyo to cut its way through the rest of the valley to the mission, more than eight miles from Hughes's ditch, swallowing Silver Lake and the Spring Branch in its course. With the completion in 1913 of the West Branch cutoff dike south of Martínez Hill, the channelization of the Santa Cruz to its present configuration was in place.

The destruction of the river and loss of prime agricultural land, however, has one bright side. The original path of the river, today's West Branch, essentially cut off and ignored for a hundred years, has accumulated the usual urban flotsam of worn tires, shopping carts, and stained mattresses. But the forgotten drainage has also become a sanctuary for rare and important animals like narrow-mouthed toads and giant spotted whiptail lizards, species found nowhere else in the Tucson Basin.

My eyes catch every movement in the tall dry grasses as I tune myself to the possibility of seeing these herptiles. A pair of Abert's towhees flee before me, darting low across the ground to find refuge in a barricade of desert broom. The season's first round-tailed ground squirrels chase each other among their dirt burrows. In the deepest cut, I step over drying mud, the irregular polygons crunching like Hershey bars.

The West Branch "officially" runs roughly seven miles from Los Reales Road to its confluence with the Santa Cruz near Sentinel Peak, most of it a deep brushy ditch clotted with desert broom and paloverde and mesquite trees among the more than 168 native plants. Here, near Valencia Road, the river is a straight channel, ten feet deep and sixty feet wide. On the west, a barbed-wire fence cordons off a stump-remnant mesquite bosque of the O'odham Nation.

It's a common misconception, often repeated by various sources, that 90-95 percent of riparian areas in the Southwest have been lost. Raymond Turner, a retired botanist for the U.S. Geological Survey on Tumamoc Hill, attributes this error to a single scientific paper about cottonwood trees on the lower Colorado River. Turner's research, however, based on comparing historical photographs of river gauging stations with modern photographs of the identical sites, leads to a different conclusion. "With some notable exceptions," he says, "riparian vegetation as viewed through the lenses of our photographers has increased at more camera stations than it has decreased."

It seems, according to Turner, that riparian loss is a myth in Arizona, except perhaps for parts of a few rivers near areas of heavy agriculture and groundwater pumping, places like this stretch of the Santa Cruz.

Historically, water flowed year-round in the West Branch, an occurrence that supported lush riparian habitat of cottonwood and willow and a dense mesquite bosque. Aerial photographs of the channel from 1936 show wide bands of mesquites, the trees contouring the river on its west bank among scattered agricultural fields. Today, relics of these giants still lift green canopies of shade above trunks measuring three feet across. Riparian loss even here isn't as great as one might think.

I step among four-wing saltbush and graythorn, which grip the eroding banks of gullies angling in from the southwest. Probably, this part of the West Branch was channelized to protect crops from storm water coming off Black Mountain and the O'odham Reservation, and then later reinforced to keep floods from inundating a growing residential area, which, as I approach Drexel Road, sprawls on both sides of the ditch. A concrete and metal footbridge spans the wide channel.

At Irvington Road, between four piers of volcanic stone top-planted with agave, two gigantic Gila monsters face opposite directions in the

West Branch channel of the Santa Cruz River

median of the road, their tails touching at the exact center of the bridge. At fifty-three feet long, including their extended, mirror-encrusted tongues, the colorful mosaic lizards are visible from aircraft on approach to Tucson International Airport. The reptiles are the work of architect Robert Vint, who worked on the exterior renovation at the San Xavier Mission. With a team of masons and tile setters, he created the sculptures using sprayed concrete, or gunite, and embedded it with broken ceramic tiles. The Mexican Tile Company provided the tiles, and their maroon, yellows, turquoise, terra-cotta, blues, and black reflect the banded pattern of the Gila monster's skin. Cup handles, marbles, bottle necks, and candle holders give the skin its beaded texture.

The Midvale Park neighborhood requested the artwork, and during its construction in 1993, people began bringing all kinds of trinkets—seashells, glass flowers, china teacups, mahjong tiles, billiard balls, bike reflectors, even a polyurethane scorpion paperweight and a ceramic statuette of the Virgin—that they found meaningful. The artists, in turn, incorporated these into the sculpture, making it truly a work of public art. Bob Vint says he wants people, when they see the twin Gila monsters glistening in the sun at Irvington bridge, to remember what a wonderful environment we live in "and to celebrate it."

Pigeons *woof* at me from their concrete perches as I pass beneath the bridge. I look for but don't see any bats. Other wildlife, however, comes to me in the presence of birds: towhees, a Cooper's and red-tailed hawk, and a first-of-the-season ash-throated flycatcher. And now, I'm being followed. I climb out of a 1983 gunite-walled, floodwater diversion channel, flushing a covey of Gambel's quail, to search for the West Branch, which I've now lost. A mockingbird sings his full repertoire of mimics from a paloverde tree, competing with the drone of traffic at the Irvington–Mission Road intersection. Among the low swales of saltbush, wildflowers spray the ground with color: gold poppy, blue lupine, the rich purples of phacelia, and bright yellows of bladderpod. From behind me I hear the clacking of a roadrunner, and then a second, but I'm not paying attention, being mesmerized by flowers. And then I feel the eyes, the head-cocked stare of something reptilian. When I turn around, two roadrunners stand looking at me only ten feet away.

"Hey," I say. "What do you guys want?" I walk on and they follow. "This is creepy," I say. "I don't have anything."

I drop my pack and open it to prove it and find two Clif Bars. I pull sticky chunks from the "chocolate brownie" flavored bar and one of the birds closes to five feet, launching himself onto my offering and then quickly rejecting it, stropping his beak against the ground.

"Fine. You're so picky," I say as his partner wanders off, no longer interested.

I tear the wrapping off the second package and toss pieces of "Black Cherry Almond," which the remaining bird seems to appreciate more.

"You're learning bad habits from pigeons," I say, finishing off the bar myself while shouldering my pack.

I'm reminded that roadrunners rule the yard where my family lives not far from here along the San Juan Wash. I've seen them swoop upon speeding whiptail lizards and dodge the strikes of diamondbacks, wearing the rattlesnakes down to the point where they can hardly lift their triangular heads and then killing them by cracking them like a whip. I've watched roadrunners devour baby quail. Nature red in beak and claw.

But life and death come with the territory. Javelina destroy our gardens, kissing bugs suck our blood, great horned owls and coyotes prey on our pets. Over the decades, we've experienced a slow attrition of pets, especially my daughters' cats, due to our wild surroundings. First Misty (whose bloody collar was all we found), then Lucky (who wasn't), then Periwinkle, Sheba, Mittens—until one neutered gray male remained: Melissa's cat, RainCloud, decapitator of baby cottontails. RainCloud has learned to live with the desert; he's a survivor. In fact, Kasondra tells a story about being in the yard once when RainCloud suddenly came streaking toward her through the bursage and prickly pear with a coyote on his tail. When the coyote looked up and spotted Kasondra, it veered away. RainCloud shot through the cat door and disappeared under Melissa's bed.

Every evening, Melissa calls RainCloud in, believing that, by locking him inside, her cat will be safe. She's trained him well. At dusk, you can hear her a quarter mile away: "RaaaaaaaaaainCloud! Diiiiiiiiiner!" And nearly every time he comes charging in, meowing for his Friskies Mixed Grill or Salmon Delight. (I can smell the opened cans as I write this!) RainCloud quickly learned that if he could get back outside before someone shut the cat door, he could get a second dinner.

Occasionally, he misses his dinner appointment—usually because he's just snagged a fat pack rat or rabbit and has decided to pass on Mixed Grill for warm brains. (RainCloud, like most cats I imagine, employs a rational methodology when feasting. The head and brains go first, leaving a bloody hole at the neck. He then reverses himself, gnawing on the haunches and working his way toward the missing head. Chewing pressure forces out the distasteful intestines, and soon all you have on your lawn is a pile of guts—something cold and slick to step on barefoot when you go for the morning paper.)

One rainy night, however, RainCloud never came in. Or the next day, and the next after that. It appeared he was gone. After several weeks went by without any sign of him, we figured the worst.

Six months later, I received a call from a woman. "I found a gray cat with a name and this number on his collar . . ." she said.

"RainCloud?" I asked.

"Yes! He's been living under our building. I noticed him several days ago, and that he had a collar. I coaxed him into the office, but he's really upset."

I recognized RainCloud's terror-stricken yowls in the background. I could hardly believe it when she gave me her address. Somehow, RainCloud made it halfway across Tucson.

I took note of Melissa's karate school as we passed it, and how close it was to the address. How could RainCloud have come this far, crossing six lanes of freeway? And why would he? I considered the possibility that he had escaped from a catnapper, but this didn't seem likely. RainCloud normally ran from strangers. I drove Melissa to karate twice a week, but surely we would have noticed him in the car as we drove the eight miles to Stone Avenue.

"He's behind the couch," the young woman told me after I introduced myself. She was a receptionist for a small tax preparation business and had been feeding several cats living in the crawl space beneath the building. "He was the only one that looked like he belonged to somebody," she said.

But when I looked behind the couch, I didn't recognize the animal staring back at me. He was dark gray, much darker than the RainCloud I knew, and he was lean and wild-eyed. But apparently he recognized me and allowed me to lift him into my arms. I released him into the car, where he hid under the seats all the way home. Once I carried him inside and locked the cat door, he disappeared under Melissa's bed for most of the day. When Melissa arrived home from school, she, too, didn't recognize her own cat. "RainCloud?" she said when she saw the strange cat on her bed. "RainCloud! Where have you been?"

That had been the salient question: Where? But after RainCloud's return home, the question became "How?"

I've since imagined that perhaps he traveled the river. It would more than double the distance, but he could have traced the wash behind our house to the Santa Cruz via the West Branch. From there, he could follow the river downstream to the Rillito River confluence, then head up the Rillito, passing beneath both I-10 and Oracle Road to Stone Avenue.

Later, Melissa returned to the woman's business with a plateful of chocolate chip cookies and a letter of thanks. She also brought a picture of herself hugging RainCloud when the two of them were both just kittens.

RainCloud's only physical reminder from his adventure is a white patch of fur on his rump. It's his badge of honor.

These days, RainCloud sleeps on my bed, white belly up, legs splayed in wild abandon. No doubt he's dreaming about his days of freedom. A smile of unmerited pleasure curls across his lips. He is so pathetic. So deplorable. So happy to be home.

I still can't locate the West Branch. I circumnavigate another new housing development with its rock-lined drainage ditches. And then I run into fences. I try to get around one by taking a narrow passage between rows of houses, but I'm blocked by more fences and the tall, oleandered perimeter of a trailer park. I backtrack and try the other way but again houses block my progress. A yard full of dogs jumps at a fence, startling me.

"Never mind!" I yell at the swirling, barking canines and reverse myself again.

A half hour later I relocate the riverbed, now a deep channel overgrown with mature mesquite and Mexican paloverde. A fan palm grows out of the bank, a volunteer from the landscaping of the Rincon Country West RV and Trailer Park, no doubt. I walk below street level, passing an abandoned and overturned shopping cart, my path a gravel riverbed ten feet wide. The air is cool, still. Blooming silverbells speckle the banks, lifting tiny, shining urns above leafy mustards. Mourning doves coo from arching mesquite branches that screen out the sky. This is the dry river I've been hoping for! Home for the homeless, and habitat of whiptails and toads.

Where the West Branch cuts toward and then parallels Mission Road, giant mesquite trunks lie on their sides, uprooted by erosion, their heavy branches sawn away many years ago. Others, however, have recovered from the wood harvesting, sprouting wickedly tangled canopies as if they were mown weeds regrown.

I stop to photograph the trees and get hung up among their thorny branches. Then I notice that directly across the road, three fifty-five-gallon metal drums, welded end-to-end and painted red and white, mark the front of Billy's Barber Shop. A sign claims the shop employs not one or two but *three* barbers. The lilac-hued construction is one of three roadside attractions on Mission Road that most locals drive past without a second thought.

Next to Billy's is Southwest Etc., a collectibles shop of western art, furniture, glassware, and other curios and doodads including life-size photo cutouts of John Wayne.

Turquoise Skies, however, the third building in line, seems to be today's tourist destination, based on the number of cars out front. Karen and I recently visited with Marshall Kidder, proprietor of the shop, which features authentic Native American jewelry, baskets, pottery, and rugs. "No drinks, food, or firearms," the sign at the entrance says, a policy apparently strictly enforced by the hand-carved and -painted wooden Indians in full headdress standing on either side of the door.

"I buy only original art," said Marshall, a man as big as his name with silver hair and goatee. "And each piece is purchased individually." Marshall has been in business here for twenty-five years, but Karen says she remembers the place when she was a child. While Karen perused some turquoise and gold rings, I looked over Tohono O'odham devil's claw baskets, Navajo sand paintings, and Mata Ortiz pottery. Eventually, I got around to mentioning to Marshall that I'm writing about the Santa Cruz River, which I pointed out flowed right past his place.

"Oh, yeah," he agreed. "And one time it flooded my shop. We had water all through here. There used to be some beautiful mesquite trees here along Mission, but the highway department bulldozed them when they repaired the road."

Karen didn't buy the rings, although she still thinks about it. She did, however, find a "baby" roadrunner metal sculpture, which now rocks back and forth in our garden beside its two "parent" birds. The birds have a cockeyed look, and I'm sure they're begging for Clif Bars.

On a hot afternoon in late May, Karen drops me off at the West Branch where it passes beneath Ajo Way and cuts along the east side of a shopping center. My plan is to follow the drainage north, back to the main stem of the Santa Cruz River near Silverlake Road, then walk along the river path to Grant Road about six miles from here.

I'm immediately impressed by the large mesquite trees hemming first the east bank and then the west. Roadway sounds drop with the temperature. Where Kennedy Park Wash drains into the West Branch, a phainopepla calls from thorny Mexican crucillo, and mockingbirds, curve-billed thrashers, and lesser goldfinches warble, tweet, and whistle among the saltbushes and paloverde. A whiptail lizard shoots through the dry grasses, too fast to be the spotted variety, I think, while a zebra-tailed lizard pauses on a dirt mound to wag its black and white racing flag. The place is alive with wildlife.

Kennedy Park Wash empties from a human-made lake half a mile to the west that is the modern equivalent of the historic and vanished Silver Lake, although today Kennedy Lake, at ten acres, is only a fraction of what Silver Lake once was. Still, the lake is a boating and fishing hot spot. Karen's brother Dan wet his first line in its warm, green waters, honing his fishing skills with the lake's finicky bluegill and channel cats. And it was here at Kennedy Lake that my daughter Kasondra first demonstrated her fishing prowess.

She was ten years old, and the fishing trip to Kennedy Lake with her sisters and me was her idea. Kasondra even did the research, calling her Uncle Dan to get the necessary information—best fishing spot, appropriate bait and test, best fishing technique. She knew what she wanted, setting the bar high: channel catfish.

The four of us arrived about midmorning after stopping at a bait shop for worms and a grocery store for Kasondra's chicken liver and corn. She led the way, taking us around the lake with our poles bobbing and buckets swinging. Other fishermen already had lines in the water, but she ignored them, passing them by and finally choosing a place near the southeast arc where a jumble of rocks rose from the shallow shoreline. There, she baited her hook, creating a shish kebab of liver and corn kernels, and I helped her cast her line in the direction she pointed. She then took a seat in the sand, gripping her pole with both hands.

I turned my attention to Melissa and Jessica, helping them untangle their lines and properly pinch off and slip Canadian crawlers onto their number 8 snelled Eagle Claw. I had no sooner cast the first line when Kasondra began calling me. "Dad!" she said. "I have something!"

She was standing, looking at me, her pole bent halfway over. I thought: *She has a snag.*

I finished helping her sisters cast their lines and went to her side. She reeled furiously but her drag only whined. Definitely a snag—but she insisted that she *felt* something. When I took her pole to adjust her drag, I pulled on the line. There *was* something! "Kasondra!" I said. "You have a fish!" I gave her back her pole and together we began pulling and reeling, pulling and reeling. And then the fish slapped the surface. "My

God, Kasondra," I said. "It's huge!" Melissa and Jessica dropped their poles and ran to us. Other fishermen watched from their folding chairs. When we finally lifted the eight-pound catfish out of the water, one of the fishermen (who had probably been there all morning) threw down his pole in disgust. We hadn't been fishing twenty minutes.

Over the next few years, Kasondra would continue to prove her fishing expertise. There was the time on the East Fork of the Black River in Arizona's White Mountains when she dropped a line in a pillow of foam literally at my feet where I was working a stream for Apache trout. In an instant she pulled out the largest rainbow of the trip, fourteen inches.

But my favorite time happened at Rose Canyon Lake in the Santa Catalina Mountains north of Tucson. The girls and I often enjoy fishing off the dam for hatchery rainbows, chumming the water with splashes of creamed corn to lure the fish to our hooks. On this occasion, the fishing was fairly poor. Memorial Day weekend had just passed, and it appeared the lake had given up the best of the takings. All we caught were a couple of fishing poles accidentally dropped from the dam by some weekend sportsmen. Then Kasondra hooked something, which she reeled in—another fishing pole—only this one was fighting her! When she landed the pole, she began reeling in its line, which spun and circled and splashed under the weight of a nice lively trout. Kasondra, my little desert angler girl, who wears a bright pink fishing vest and catches fish in extraordinary ways.

While catfish nuzzle the muddy depths of Kennedy Lake, native fish like Gila topminnows are making a comeback in what are known as "refugia ponds" along this part of the West Branch. In fact, it is here, north of Ajo Way, that the relic river is the least degraded and has the greatest biodiversity of the Santa Cruz River floodplain near Tucson, according to Phil Rosen, a University of Arizona professor who has been studying the West Branch since 2000. It was Rosen who, in the summer of that same year, discovered the isolated but thriving population of giant spotted whiptail lizards while checking out a tip from another researcher.

"I assumed this was merely a case of whiptail lizard confusion, which has haunted American herpetology for over a century," Rosen told me when I asked him about the discovery. "To our surprise, we found the giant spotted whiptail persisting in modest abundance as a large, attractive, very alert animal in the dense thickets of the West Branch."

Giant spotted whiptails, heavy, jaguar-patterned reptiles that grow more than a foot and a half in length, normally inhabit dry, narrow canyons like Sabino Canyon. The West Branch is the only place that these whiptails live in the Tucson valley. Rosen says the presence of the whiptails is the "first and most irrefutable indicator" that the area is a unique resource. And they are only one species among ten and possibly fifteen other lizard species, a number that equals or exceeds the greatest lizard diversity in the Southwest.

Another one of Rosen's discoveries, the Sinaloan narrow-mouthed toad (actually a frog), is one of six kinds of frogs and toads known to breed along the West Branch. This is the only place within fifty miles that the chert-green, arrowheadlike amphibian occurs.

When I think about the diversity of these herptiles, then add the one hundred-plus species of birds, the diminutive black-headed snakes and "headless" blind snakes, the jewel-domed western box turtles and Christmas fields of native poinsettia, the wonderfully named Tumamoc globeberry, and desert seepweed, I begin to understand how fortunate we are to have neglected this place, and how important it is now to pay attention.

Rosen suggests that one of the first steps in caring for the West Branch is encouraging people to interface with nature. "People are already living within the natural environment," he says, "and in some cases the best habitat is on private property. Neighborhoods, not just open space, could be incorporated into the wildlife and plant population dynamics, if residents develop their vegetation, plantings, and yards with elements that fit into the broader landscape and the natural setting."

People interfacing with nature is the thrust behind the Arizona Open Land Trust's efforts at the West Branch. Since 1978, the trust's mission has included protecting Arizona's western landscapes and wildlife through land stewardship, legislation, public education, and outreach. The trust became involved with the West Branch in 2000 when area neighbors expressed concerns about Pima County locating a bus barn near the channel. After Phil Rosen conducted a natural resource inventory of the West Branch, people realized they needed a plan to protect both the environment and their agrarian culture/urban lifestyle in light of the forthcoming Rio Nuevo Project, Tucson's controversial but important effort to breathe new life into the city's downtown, including this reach of the Santa Cruz River.

With help from the Arizona Open Land Trust and the U.S. Fish and Wildlife Partners Program, some residents dug refugia ponds to help reintroduce lowland leopard frogs; already they have seen the appearance of six kinds of toads. Other property owners encouraged the regeneration of mesquite and blue paloverde, creating one of the highest quality xeroriparian bosques in the city. This habitat draws an amazing diversity of turtles and snakes and amphibians, and the ponds have brought the return of native fish, something unheard of in the Santa Cruz River until now.

Today, the trust is supporting—from these backyard pond projects to outright land purchases—the ongoing protection of the whole western watershed of the West Branch, from the channel itself to the Tucson Mountains, an area encompassing roughly ten square miles and ten major washes, including the San Juan outside my back door.

I'm already expecting the sheep-bleats of breeding Couch's spadefoots on quiet evenings following a summer thunderstorm. Maybe one day I'll get to know the buzzy-whistling of the Sonoran green toad, the reticulated "troll of exceptional beauty" that Rosen wants to reintroduce here.

I recently visited with two residents of the West Branch, Janine and Tim McCabe, who live along "my" San Juan Wash where it passes under Mission Road and crosses the property of Judy Fraser, a feisty woman with salt-and-pepper hair who has ranched here since the early 1960s. Tim and Janine have been renting Judy's guesthouse going on twelve years.

Janine, a lithe, blonde, thoughtful woman, works at a plant nursery and leads bird-watching field trips for the Tucson Audubon Society. She wants to encourage more people to visit the West Branch, and thinks the area should be included in a future edition of *Finding Birds in Southeastern Arizona*, a Tucson Audubon Society publication. "We have great wintering sparrow habitat most years," Janine told me. "And a real good migratory pathway." Among the bird sightings she mentions are nesting Bell's vireos and black-crowned night-heron and greater yellowlegs flyovers. Recently, she added a yellow-breasted chat to her "yard list," which now stands at 110 birds, an impressive number for any one location.

Janine and Tim gave me a tour of what the residents want to preserve, and while Tim and I chatted about his work designing and building trails with Pima County Parks and Recreation (he helped create a trail I hike most days at the end of my street), Janine led us through an ancient scrapyard of weathered lumber, irrigation pipes, and rusting ranch equipment to the area's first refugia pond. In the midst of the junk, it appeared like a small oasis, a black pool bordered by flowering bulrushes.

"Some of the fish are pretty large," Janine said, referring to the Gila topminnows that darted just beneath the flat surface. The last time I'd seen these native fish on the river was at Sonoita Creek near Rio Rico, one of the last natural refuges for the guppy look-alike. "This is the pond where Rosen also introduced the lowland leopard frogs," Janine added.

We continued through a ditch-bosque, past a "tub-pond" filled with bright green duckweed, which Janine explained that Judy Fraser was maintaining for narrow-mouthed toads, one of the signature species of the West Branch. I remembered that just opposite the refugia ponds, across the Santa Cruz River half a mile to the east, a desert pool called the "Mesquite Circle Pond" is home to the greatest diversity of amphibians and the best-known breeding site for Great Plains toads in Tucson. In 2006, researchers recorded the vocalizations of narrow-mouthed toads there, the first ever recorded in the valley east of the Santa Cruz River.

Last March, I walked that wallow of dried and cracked mud, its margins green with peppergrass under skeletal mesquite branches, the way it appears most of the year. But then, usually in July, monsoon thunderstorms will fill the depression with rainwater and toads will rise to the surface in near-spontaneous droves.

I had rising toads on my mind as we circled back past Beryl Baker's property, her mewing sheep, and her own refugia pond lined with smooth river stones. That day the only amphibian we found was a large desiccated Sonoran desert toad, its body still olive-green and perfect, though a bit shriveled. Tim thought the toad had been dead for some time, but Janine wondered why she hadn't seen it before; nothing here escapes her notice—not the distant lurking coyote, the diminutive vireos, or the recalcitrant jackrabbit, which she sees as a symbol for the West Branch's recovery. I considered pouring water over the animal, and I wouldn't have been surprised if the amphibian instantly began to swell, stretched out its chubby legs, and then casually hopped away as if nothing unusual had happened to it. This place makes me believe in resurrection.

I leave the sandy drainage southeast of the ponds, and step into an open flatland ringed with mesquite and furred with dead amaranth. The empty sockets of a calf skull stare at me from a trail through the prickly stalks. Where a wide and shallow Enchanted Hills Wash crosses the floodplain from the west, I enter the longest-standing, tallest, and densest mesquite bosque on the West Branch. Floods in the 1970s filled in and eliminated the wash's deep channel, creating a wide delta for occasional runoff to sheet flow, which in turn allowed this impressive regeneration of mesquite and blue paloverde with its thick understory of saltbush and acacia shrubs. This is the best dry river habitat in the city, and I have trouble navigating it to return to the West Branch with its flocks of squeaky-winged mourning doves.

Fifteen minutes later, I arrive at what I'm certain is San Juan Wash. I'm struck by the emotion I feel, being so simply connected to my home, only a mile away. Even the sand beneath my feet carries a special significance. How long ago did I or Karen or one of our three girls stand on this same earth? I look for something I might recognize—stranded Pooh sticks, a lost basketball, maybe a forgotten shoe—and realize I recognize it all.

I'm not surprised when my cell phone rings and it's Melissa. "Guess where I am," I say.

Melissa is working today at the University of Arizona's Science Operations Center, which houses the Phoenix Mars Scout Mission. There, in the Payload Interoperability Testbed (the "PIT"), researchers practice on a full-scale engineering model of the lander before they send commands to the craft on Mars. "We have photos from HiRISE showing Phoenix with chutes open descending to the surface!" she tells me.

Yesterday, Karen and I joined Melissa and several hundred scientists, their families, and media people in the PIT for the historic landing. Now moving into her fourth year at the U of A, Melissa has been involved with the Lunar and Planetary Lab (LPL) since high school. In fact, her senior year science fair project lined her up perfectly with this current Mars mission. After the previous year's project when she successfully grew Antarctic bacteria in Martian-like temperatures and soil, she decided to raise the bar. This time she would attempt to grow bacteria in complete Martian conditions, not just soil and temperature but atmospheric as well—in the bitterly cold, dry, carbon-dioxide-laden near-vacuum of the polar regions. And not just any bacteria, but those germs she would swab from the Phoenix lander itself.

She didn't require me to dig holes in the desert for this experiment. She wanted "Martian chambers." What we came up with, after abandoning the offerings from a scientific equipment store and turning instead to a plumbing supply shop, were three heavy, eight-inch metal containers that, except for the valves and pressure gauges sticking out of them, looked a lot like pipe bombs.

The bacteria from the lander bloomed in her chambers, better even than the bacteria from Antarctica, which got the attention of LPL scientists working on the Phoenix mission. Melissa's project began focusing on "planetary protection," a concept NASA officials had recently become interested in with its Mars missions. Melissa started repeating a citation from her research, saying that if life is discovered on Mars, we should pick up our trash.

Since her first years at the U of A, Melissa has worked with the Phoenix mission, sitting in the clean room calibrating the camera on the robotic arm with its digging tool and lately visiting schools and giving tours in the PIT as the mission's educational coordinator. (I tell people that she could be the next Carl Sagan.) She has also become, with the help of a NASA space grant, the LPL's "historian," interviewing dozens of scientists involved with the lab since the 1960s and recording their stories.

Karen and I met several of these scientists yesterday, as they hugged our daughter and swung her around the room in celebration of what the LPL has aimed at for decades—the complete management of a spacecraft. It would begin as soon as NASA's Jet Propulsion Laboratory (JPL) in Pasadena delivered Phoenix to the Red Planet.

At 5:39 p.m., the first indication of Phoenix's landing came as JPL engineer Richard Kornfeld relayed to the crowd that the cruise stage separation had occurred. One screen began to show a computer simulation of the landing, while another pictured the climbing red graph of a Doppler radar. The room fell silent as he announced atmospheric entry and peak heating, the beginning of the so-called seven minutes of terror.

Minutes went by.

Then, "parachute detected," followed by a burst of cheers and "we have data lockup." People were holding their breath, holding on to each other. When I turned to look at Melissa, she was holding the fingers of both hands between her teeth. This was the point at which the previous Mars polar lander had failed, crashing to the frozen surface of the southern polar region on December 3, 1999. The tension grew as Richard called out the lander's separation from the backshell and chute, the throttle-up of its engines, and then he began reading elevations: "six hundred meters . . . five hundred . . . two-fifty . . . eighty meters . . . sixty meters . . . forty meters . . . twenty . . . Stand by for touchdown signal . . ."

Then silence, a very long silence, until suddenly, "Touchdown signal detected!"

The room erupted. People had tears streaming down their faces. I wrapped my arms around Melissa, saying, "Your spaceship is on Mars!"

Today, I still choke up thinking about it, feeling the weakness in my knees. Melissa has been cruising on a trajectory in her life as certain as the path of her spacecraft. What began eight years ago with digging holes in the dirt of the Santa Cruz River has led to digging holes in the red dirt of Mars.

The West Branch deepens, the surrounding floodplain rising to twenty feet on either side of me as I begin to approach the drainage's end. I search the ocher walls of both sheer banks for the telltale dark line of lake-bottom sediments, thinking there still might be some evidence of the large shining body of water.

In the late 1850s, Alfred and William Rowlett stretched a low earthen dam across this once-marshy river floodplain and constructed Tucson's first flour mill west of the resulting lake. Using the flow from several springs to drive its water wheel, the mill then emptied its tailrace into the lake, from which a public acequia irrigated farmland to the north.

In 1860, a Maine businessman and military contractor named William S. Grant bought the brothers' mill and reconstructed it along with the dam. Flooding seems to have been a problem for the dam and mill, but Grant must have been equally frustrated with the high tides of war. Union soldiers would destroy the mill in 1861—along with Grant's Tucson warehouse and all of its supplies—while retreating before Confederate forces (no bread for Johnny Rebel) and then, a year later, rebuild it upon their return.

By 1864, the Irish teamster and miner James Lee arrived on the scene, rebuilding (again) the Rowlett brothers' mill. Like many other Tucsonans at the time, Lee took his wife and family to Mexico during the Confederate occupancy, but he soon returned to partner with William Scott and start a milling business on the Santa Cruz River.

His was an illustrious career. When Lee wasn't grinding fresh-harvested wheat or dragging a steam engine and boiler into the Santa Rita Mountains for his lumber mill, or digging five hundred pounds of silver ore from his mine, La Naguilla, he was investigating criminal activities and tracking stolen livestock into Mexico to shoot it out with rustlers. Twice he ran for sheriff. Later, he became captain of the Arizona Minute Men, a group of citizens formed to protect people from "thieves and murderers." Before he died from pneumonia in 1884, Lee had discovered a "rich black body of sulphurate ore" near Oracle in the Catalina Mountains and opened the Pilot silver mine. Although he's mostly remembered for his Pioneer Mill at Silver Lake, James Lee provided Tucson with much more than white flour for the two decades he lived beside his pond with his wife María and their nine children.

Sometime in the 1880s, "Lee's Pond" adopted the more glamorous name of "Silver Lake" after Lee leased part of it to J. F. Richey and J. O. Bailey, and later Frederick Maish and Thomas Driscoll, fresh from their Canoa Canal Company project to bring water to Tucson. The developers built a hotel and bathhouses, pavilions with shaded cottonwood groves, and even a mile-long racetrack, tying the recreational spot to the city with a tram service. Then, as the beautiful resort grew in popularity, Tucson's first civil engineer, U.S. Surveyor George J. Roskruge opened it to sailboating, swimming, and picnicking.

All of this is gone now. The lake and trees, the hotel and bathhouses. The racetrack is beneath Cottonwood Lane, now a paved road just west of the river. All are replaced by a wide trench that never really was the Santa Cruz, a trench described in 1890 by a bitter George Roskruge as "Sam Hughes' ditch taking a walk to [Silver Lake] after water."

The only indication that there was ever a lake here comes from the street sign: "Silverlake."

I scramble up a steep slope of crumbling soil cement and stand on the bridge that crosses over the present-day West Branch where it commingles its sediments with those of the Santa Cruz River. The confluence is a human-made reconfiguration from the 1960s that shifted the West Branch from its original course under Sentinel Peak. From the bridge and bike path, I can see where the Santa Cruz's broad swath of sand sweeps in from the south, the channel a deep cut with gray-walled banks.

I remember what Phil Rosen said about the West Branch, how the whole system could be brought back if we work to understand its conservation and look for opportunities for its restoration. What's most encouraging to me is that it's already begun. A plan called Paseo de las

Iglesias, "walk of the churches," outlines how city, county, and federal agencies are joining together with citizens to recover the river's flow and its riparian habitats between the San Xavier Mission and the soon-to-be reconstructed mission of San Agustín del Tucson at the new Tucson Origins Heritage Park. By planting a thousand acres of mesquite trees, saltbush, wolfberry, and brittlebush along with twenty acres of cottonwoods, willows, and bulrushes, the project will enhance wildlife habitat along the corridor, including completing the missing trail links to the river park as part of the Juan Bautista de Anza National Trail.

I imagine walking this path between the missions with cottonwoods flashing their leaves in a cool morning breeze. Sunlight shimmers in a slow slip of backwater beneath them. Black phoebes and vermilion flycatchers chase insects from streamside snags, and from somewhere among the thick emerald canopies a gray hawk screams.

For a hundred years or more the West Branch, essentially the historical Santa Cruz River and the reason for Tucson's existence, has been waiting in its neglect, waiting to be noticed and revived, waiting with much of its original natural character intact. Now, in this new millennium, the river may finally rise again.

"The West Branch retains the history of the Santa Cruz—the Sonoran Mexican, south-facing history," Phil Rosen says. "I don't think an animal like the Sinaloan narrow-mouthed toad ever lived north of downtown. It is on the West Branch today because that's where it was one hundred and one thousand years ago."

Sentinel Peak, 121 River Miles

An hour and a half after Karen drops me at Ajo Way, I slip under the 22nd Street/Starr Pass bridge on the west bank of the Santa Cruz River. A dozen northern rough-winged swallows dip and circle around my head, blowing lipless raspberries at me. On the east bank opposite me, hundreds of thriving native plants crowd a quiet shaded enclosure of chain-link. Like the West Branch itself, the green three-acre enclave is a "species bank," only this one is human-made.

Desert Survivors Native Plant Nursery is part of a human service agency started by a Tucson psychologist. In the 1970s, Dr. Joseph Patterson recognized that his patients' needs for accomplishment and self-esteem could be met by the kind of physical work and wholesome environment plant nurseries provide—in essence, he understood the healing power of green plants, soil, and sun. In 1981, he founded Desert Survivors, Inc., bringing together the desert and people with disabilities. Today, the plant nursery continues to reach out to adults with special needs by giving them meaningful employment and vocational training. The operation has been hugely successful in its combined human/environmental ethic.

The nursery features over four hundred species of native plants that live within five hundred miles of Tucson, including desert specialties indigenous to the Sonoran, Chihuahuan, and Mojave deserts. Nursery workers believe that native, regional plants give Tucson a sense of place and are important for wildlife like hummingbirds, moths, and butterflies, many of which you can see in their pollinator demonstration garden. The nursery also strives to be ecologically sensitive by using only organic gardening practices, part of what its people call their "go-native" planting philosophy.

Recently, Desert Survivors partnered with the Kino Heritage Fruit Trees Project to locate, propagate, and reestablish historical fruit trees to the original orchards and gardens at Tumacácori National Historical Park and the new Tucson Origins Heritage Park. Project researchers have begun reviewing Father Kino's journals and other early accounts, while visiting mission communities, ranches, and abandoned farms in Sonora as well as the backyards of nineteenth-century homes in southern Arizona, searching for colonial-period trees. The idea is to identify stocks introduced in the eighteenth and nineteenth centuries—trees like fig, walnut, olive, apple, peach, pear, pomegranate, date, and quince—collect seeds and cuttings from them, and grow them at nurseries like Desert Survivors.

Karen and I have been buying our landscaping plants here for twenty years. Recently, while pulling around a flatbed cart to gather mesquite trees for our yard, we found a pair of mission black figs, the same kind once grown in the mission gardens right here under Sentinel Peak. The trees have already produced fruit, firm dark globes with the deliciously sweet flavor of Spanish history.

In 1904, a photographer named Walter P. Hadsell, probably working out of Henry Buehman's Tucson studio, climbed Sentinel Peak to photograph the part of the Santa Cruz Valley looking toward Tucson. It's a classic photo. One reprinted in many books and magazine/newspaper articles dealing with the history and plight of the river. It shows the place under the peak where the West Branch angles in from the southwest to join the main stem of the Santa Cruz, the north-flowing river marked by a line of winter-bared cottonwoods. Both tributaries have steep banks, and a crowd of mesquite trees entirely fills the Y formed at the confluence. Mission Road, a single dirt track picketed on one side with a barbed-wire fence, cuts along the hillside between the mountain and the river. Except for the familiar ridgeline of the Santa Catalina Mountains across the upper frame, it could be a picture of the modern, cottonwood-shrouded San Pedro River.

A hundred and four years after Hadsell dragged his camera up Sentinel Peak, I scale the same rocky slopes to line up his vantage point with my own camera lens. I find it fairly easily, taking the 22nd Street/Starr Pass overpass above Mission Road and then following trails made by the homeless.

A remnant of an earthen dam appears in the 1904 photograph, which Hadsell took fourteen years after the Tucson Arroyo entrenchment. The washed-out structure rests just above the former confluence of the West Branch and the Santa Cruz. Today, this is where a paved, two-lane Mission Road jogs to the right going north along the side of the mountain. Some installation art showing yellow horned lizards scaling a black rock wall stands where the quarter-mile-wide dam once held back the waters of the West Branch in a fifty-acre impoundment known as Warner's Lake.

Solomon Warner was a gray-bearded double of Ulysses S. Grant whose barrel-sized chest resonated with his deep voice. He came to Tucson in early 1856 at the same time as James Lee, leading a thirteen-mule train loaded with California goods to open a small mercantile on Main Street inside the Presidio walls. It would be Tucson's first "American" store. Refusing to swear allegiance to the Confederacy during Tucson's rebel occupation in 1862, Warner fled to Santa Cruz, Sonora, where he married a Mexican widow, started another store and freight business, and lived like an elite aristocrat. After the war, he returned to Tucson and began buying properties. One of these was a parcel of land at the base of Sentinel Peak.

Under Sentinel Peak across from the intersection of Mission Road and Mission Lane, I climb among some old ruined walls of local stone—all that remains of Solomon Warner's water-powered flour and ore-processing mill after the two-story structure was dynamited for safety reasons in the 1930s. I carry an 1880 photograph by Carlton Watkins of the irregular polygons of Tucson's agricultural fields taken from Sentinel

Historical photo of the Santa Cruz at Sentinel Peak

Modern photo of the Santa Cruz at Sentinel Peak

Solomon Warner's mill site

Peak. The yellowed image barely captures Warner's Mill in the far bottom left corner, where the square structure is tucked into a small copse of trees in an otherwise flat bottomland and indistinguishable Santa Cruz River.

Warner built this mill in 1874–75, its waterwheel and granite millstone turned by six hundred cubic feet of water under the pressure of an eleven-foot drop from its source at Lee's Pond. A millrace, part of which I can still see today as low parallel walls of cemented stones on the south side of the ruins, carried the water the mile or so distance. A tailrace sent it along Mission Lane past the old Convento and into an acequia for irrigating farmland.

Three years later, Warner added a three-stamp mill for crushing ore. But by the early 1880s his power source began to dry up as water levels fell at Lee's Pond due to the Tucson Water Company's new headcut south of the lake near Valencia Road. The private operation had begun piping water to the city via a six-mile redwood flume.

Warner responded by taking advantage of a separate water source, that coming from the Sierrita Mountains by way of the West Branch as well as several springs flowing out from under Sentinel Peak. Warner's dam not only ensured his mill operations but created a local sensation for hunters, a club of which obtained exclusive permission to shoot the lake's waterfowl.

Solomon Warner closed his mill sometime around 1891, no doubt due to the devastation caused by the floods of 1890, the entrenchment of the Santa Cruz River, and loss of his dam. Perhaps he figured it was just as well. An act of God. The frustration that his mill wasn't much of a financial success, and the many long years of fighting for water, the lawsuits from the builder of his millraces and from the downstream farmers—all of this would have worn anyone down.

All the same, I find it intriguing that, although these few stone walls are all that remain of his mill, his house still stands where it has for more than 130 years. Warner's home has outlasted Warner himself. I imagine him finishing his years in this place, living, as the historian C. L. Sonnichsen says, "in less-than-genteel poverty working on his perpetual-motion machine."

Today, Diana Hadley and Peter Warshall live in Solomon Warner's home. Diana is a historian and associate curator of ethnohistory at the Arizona State Museum. She has published numerous studies on the history of land use and ecological change in the Southwest and northern Sonora, and she's one of the incorporators of the Northern Jaguar Project, where she serves as its president. Peter, the project's secretary, has spent thirty years working with conservation groups, businesses, and governments in places like Africa, Latin America, and the United States on biodiversity protection. He's known as an expert on our sky islands and serves on Sky Island Alliance's board of directors.

I first met Peter at his home during a Sky Island Alliance fund-raiser a couple of years ago. As he learned about my travels on and writing about the Santa Cruz River, he became more and more animated, leading me through his house, his long silver hair flying, talking about Solomon Warner and the restoration he and Diana were working on, showing me historic photographs of the mill site and the Santa Cruz Valley hanging on his walls.

"We are so fortunate to live here," he told me, "under this mountain where Tucson began."

Hohokam mortero at Sentinel Peak with a survey marker

Behind Warner's Mill, I scramble along this steep rocky flank of Sentinel Peak, dodging cholla cacti and the flesh-clutching whips of acacia. Suddenly, I'm staring at three large morteros, Hohokam grinding holes for corn, embedded in low black rock. The holes, more than a foot deep and tapering into rounded bowls, lie just above Warner's half-buried millrace. And then I find others, one only a dozen feet from Mission Road. This is amazing, I think. Separated by half a millennium, Warner chose the same site as the original people to grind his grain into flour.

In the afternoon light, Hohokam circle petroglyphs materialize on the black volcanic boulders. Other shapes also appear, each one pecked into the wide varnished surfaces, some more recent than others. The mountain is a canvas for symbols.

Sentinel Peak is so named because early settlers used the outcrop to keep watch for raiding Apaches. Today, most people know the peak as "A" Mountain for its whitewashed (or lately, its patriotic red, white, and blue) designation for the University of Arizona. Sentinel Peak's original name, based on its description, however, is what many believe gave our city its name.

In his book *Spanish Colonial Tucson: A Demographic History*, anthropologist Henry Dobyns writes one of the clearest and most comprehensive explanations of the origin of the place name that I've ever read. I won't go into all of his three paragraphs of detail, but he basically argues that "Tucson" is a Spanish derivation of *schookson* (or *stjukshon*), an O'odham word meaning "at the foot of black" (mountain). "Black Mountain," as anyone from the tribe will tell you, is Sentinel Peak.

When John Bartlett came to Tucson in July of 1852, he, like others, chose Sentinel Peak to get a perspective of the valley. He describes Tucson as adobe houses in ruins. "No attention seems to be given to repair," he says, "but as soon as a dwelling becomes uninhabitable, it is deserted, the miserable tenants creeping into some other hovel where they may eke out their existence." His panoramic drawing from the saguaro-stabbed mountain shows a view of these "hovels" with the mission of San Agustín del Tucson in the foreground. Bartlett calls the gleaming white Convento and church a "large hacienda" and writes that it must have been very rich at one time, but that "it is now in a decayed state, and but a small portion of its fine lands are cultivated."

San Agustín del Tucson, 122 River Miles

One hundred and fifty-four years before Bartlett's visit, on a warm September in 1698, the first European to arrive at what would be Tucson probably climbed this same peak to look over the valley of the Santa Cruz River. Father Eusebio Francisco Kino would have seen a green floodplain, dotted with beehive-like brush huts and lined with ditches carrying river water to fields of ripening cotton, beans, and squash. Here lay the second largest community on the northernmost frontier (after San Xavier del Bac), an O'odham village of as many as 750 inhabitants, a brown-skinned people whose digging sticks could have been handed down from the Hohokam. Kino called it San Cosmé del Tucson.

The Jesuits attempted to install someone to serve the native communities at Bac and Tucson, but San Xavier's first missionary, twenty-eight-year-old Father Francisco Gonzalvo, lasted little more than a year, dying "from a cold" in August of 1702. The padre's life at Bac was little more than an afterthought. The mission and its *visita* had no resident priest for more than thirty years. What Kino had started would be left to others to complete. It was Father Jacobo Sedelmayr who, after the Pima Revolt in November 1751, recognized the strategic significance of Tucson's location and suggested that the Spanish build a fort here. But there would still be another twenty years of strife, both internal (the 1767 Jesuit expulsion) and external (Apache warfare), before the first permanent European settlement began rising on the banks of the Santa Cruz.

In early 1771, Father Francisco Garcés, the restless, Kino-class Franciscan assigned to San Xavier after the Jesuit expulsion, constructed a massive castlelike residence with a high round tower at each of its four corners. The defenses were soon tested when Apaches attacked the community, the soldiers and citizens holding off the raiders from behind the walls and finally repelling them. The following year, perhaps encouraged by this success, Father Garcés began a mission house and an adobe chapel, which he named in honor of Saint Agustín.

As it expanded over the next few years, Father Garcés's building program revealed his ambitions for the community. His granary measured almost sixty feet by thirty feet. With soldiers from Tubac providing some security, his fortifications protected both people and the mission's livestock. The small herds of horses and cattle grazing in the valley could be gathered inside their high walls the moment a lookout on Sentinel Peak gave warning of approaching Apaches. But Garcés's boldest project must have been the *casa conventual*, the Convento, Tucson's first convention center.

Many historians believe it was Garcés who designed this "place for convening" around 1778–79, sometime after his return from the Anza expedition and before his departure for the Yuma mission, where he would die in 1781. He began by enclosing a three-thousand-square-foot area with a foundation of basalt boulders tucked into adobe mud. The Convento's five ground-floor rooms would be the administrative offices and residences for the mission priests and the center of a walled complex of one hundred thousand square feet, rivaling that of even San Xavier.

I've studied the few photographs that represent all that remains of the two-story Convento. Although in Bartlett's drawing the giant structure appears intact, the earliest photos from the 1880s show a crumbling and roofless mud ruin. Even so, it appears out of place for Spanish colonial Tucson.

Garcés's Convento would have been simpler, its single-story walls whitewashed inside and out and a roof of mud laid over saguaro ribs or ocotillo branches supported by beams of mesquite or pine. The Convento's second floor came later, probably between 1798 and 1810, under the direction of Father Juan Bautista Llorens, who would have been fresh from completing the San Xavier Mission. Llorens added more rooms, with a central arched corridor and adobe stairway below and arched windows with a covered balcony above. He set bricks in lime mortar for floors and colored the walls with red and white plaster. Then, he turned his gaze upon Garcés's little church. If Garcés was ambitious, Llorens was ostentatious.

At the height of Father Llorens's presence at San Agustín del Tucson, ten-foot walls, eighteen inches thick, would enclose the mission complex, its mesquite-wood gate open at the center of the south wall. Alongside the shining Convento, a new brick church, painted like rust, would raise a high tower holding three copper bells against a blue sky. Inside, its altar might hold chalices and candlesticks made of silver from the Tumacácori mines, its walls hung with the oil paintings of saints. But San Agustín was more than an impressive church and convento. Adobe houses and workshops, canal-irrigated gardens and orchards spread out over the area. Burning mesquite wood trailed from the soap factory and kitchen. Villagers, both native and Spanish, stripped to breechcloths and buckskin guaraches, stretched hides at the tannery or stirred liquid beeswax for candle making. The air filled with the smoke and dust of activity, the ring of the hammer and rasp of the saw.

Although Father Llorens would complete the Convento and replace the adobe church, it was Father Garcés who first had the vision for the *pueblito* at the foot of Black Mountain. It was Garcés, the gentle, wandering Franciscan whose every genuflection brought him closer to Yuma, who lifted out of the dirt the "little pueblo" of brush huts and bean plots to fashion a town that people would one day call the Old Pueblo.

From Sentinel Peak, I cross a busy Mission Road and walk along Mission Lane toward the river and Father Garcés Convento site. Small square houses surrounded with short rusted fences line the street on my left, and they make me think of what longtime resident Lillian Lopez-Grant says about this historic neighborhood: "Happiness in Menlo Park is a brick home with a chain-link fence."

People here will tell you that the neighborhood does *not* get its name from the New Jersey home of Thomas Edison. Actually, the name is attributed to Henry E. Schwalen, a chicken-rancher-turned-real-estate-developer who, in 1912, borrowed it for his new subdivision from a town he favored in the San Francisco Bay area. Schwalen's model community came with covenants that kept homes simple and created setbacks for front yards, trees, and sidewalks. It also came with deed restrictions that limited property owners to whites. The restrictions, however, were eventually ignored. Menlo Park, started as an Anglo enclave, was largely a Hispanic neighborhood by the 1940s.

Mission Lane forms the northern border of the mission gardens, a four-acre walled plot where Convento friars once tended fruit trees and vegetables. As part of the Tucson Origins Heritage Park, the city plans to reconstruct the gardens at the same location, constructing acequias and canals and planting grape arbors and orchards of mulberry, apple, apricot, fig, quince, and pomegranate. But the gardens, enclosed like the original with an eight-foot adobe wall made from the dirt beneath it, will be more than a representation of a monk's quiet retreat. The gardens will also honor Tucson's first farmers by interpreting four thousand years of Native American agriculture.

But to do this, the city has had to first deal with sixty years of trash.

Several years ago, driving along Mission Road, you would have seen a field of white pipes crisscrossing this place. Engineers were testing an aerobic composting method by pumping four hundred gallons of oxygenated water a day into the landfill and letting bacteria do the rest. After six months (and, I imagine, the release of tons of carbon dioxide and enough heat to pit-roast a small herd of cattle), the area had settled one foot.

Ray Murray, of Tucson's Environmental Management Department, called the pilot project encouraging and said it should be expanded to the surrounding five-acre site, estimating it would take two years to fully degrade all the material in the landfill. Apparently, looking at the place today, turning the landfill into a bioreactor was a success.

The Nearmont Landfill is one of three landfills in the Tucson Origins Heritage Park. One of the others, unfortunately, occupies the former location of the Convento, the remains of which the city bulldozed in the late 1950s to make way for its garbage.

The Convento's demise began as the native and Spanish population ebbed in the early 1800s. In 1814, a year before his death, Father Llorens left Bac, disillusioned by northern Sonora's political unrest and fearful of the smoldering resentment of his charges, "fed by the continuous tormenting of Father Arriquibar and his agents," Llorens wrote in a letter to his superior. (Father Pedro Antonio de Arriquibar, a corpulent Basque who left his Franciscan order for a chaplaincy at the Tucson presidio, became one of the richest men at the post, despite a Spanish decree outlawing clergy from accumulating property.) Then, flexing its new political muscles after winning its independence from Spain in 1821, Mexico sought to destroy the mission system entirely. In 1831, Father Rafael Diaz, a naturalized Mexican and one of the two priests who remained in Pimería Alta after the Mexican government expelled the Spanish in 1828, departed for the church at Ímuris. He took with him the statues of the saints, the baptismal font, holy vestments and vessels, and other religious items, leaving the Tucson mission to its slow decay and eventual end as a garbage-littered landfill. Unlike the Tumacácori mission, which nearly suffered a similar fate, there would be no 1908 national park designation for San Agustín.

I've traveled to San Jose de Ímuris in northern Sonora, where one of San Agustín's religious relics, the famous "Black Christ," Nuestro Señor de Esquipulas, hangs on a wall behind the plain altar and chairs. Originally, the crucifix was part of furnishings of the Convento, but stories say that Mexican soldiers found it in a hidden room sometime around the Gadsden Purchase and removed it to the little Sonoran church. Interestingly, someone repainted the figure of Christ so that now the image is white.

I've also visited the site of San Agustín and seen what's left of it. In November 2000, archaeologists began searching for physical evidence of the mission without much success. They did uncover the stony foundation of the complex's west wall and that of the granary, but they could find no trace of the Convento and chapel.

"There's nothing left," researchers with Desert Archaeology, Inc. told the small crowds that came to the public excavations. Most of what the archaeologists uncovered were artifacts from the 1950s, some of them in amazing condition. Homer Thiel, an archaeologist who has been working with Desert Archaeology since 1992 on projects including

Tucson's Spanish presidio, said, "We found a phone book from 1958 and looked up the number of Julian Hayden, a prominent archaeologist."

It's curious to me that, although the Mission of San Agustín had been a growing multiethnic community at Tucson for decades earlier, Tucson claims its date of birth as August 20, 1775. On that date, Colonel Hugo O'Conor, the Irish "Red Captain" inspector general of northern Mexico, ordered the Tubac presidio moved to Tucson.

Why should the military get the credit for our birth and not the church? What about that day, the 27th of September, 1698, when Father Kino climbed Sentinel Peak and named the valley before him "San Cosmé del Tucson"? Or at the very least, the day in 1768 when Father Garcés came to the Santa Cruz Valley after the Jesuit expulsion. Garcés, a Spaniard who actually risked his life to build Tucson, makes more sense as the founder of our city than an Irish mercenary who visited here for a brief military inspection. (I say this despite a rebuke from my own Irish ancestors!) O'Conor didn't even give the new post a name but ordered it called after the mission Garcés had named. Perhaps it's an "east-west" thing. Maybe our city officials continue to be averse to recognizing as relevant anything coming from the west side of the Santa Cruz River, the place of Tucson's landfills, human and otherwise.

The church and the military have always wrestled with each other. In all of New Spain the "two majesties" competed for authority and limited resources, and yet each also relied on the other. Nothing illustrates their entwined, double-helix relationship better than at Tucson where the embodiment of their presence, the mission and the presidio, were separated—and at the same time joined—by a broad meandering river that ran between them.

Presidio San Agustín del Tucson, 122 River Miles

While workers laid the adobe bricks of the Mission of San Agustín del Tucson on the west bank of the Santa Cruz, a second fortified community was rising on its east bank more than a mile away. At eleven acres, the Presidio San Agustín del Tucson would become the largest Spanish garrison in the Southwest.

During the years of its reconstruction, I often looked down from the fourth floor of a downtown Tucson office building as course upon course of adobe bricks raised a corner of our eighteenth-century presidio. At the intersection of Church and Washington, where the northeast walls of the original fort once stood, I watched as a three-acre parking lot became a historic place.

My interest began at 10:00 a.m. on December 10, 2002, as I walked along Church Street on my way to meet Karen at the Transamerica Building where she worked. Archaeologists had fenced off the lot and peeled away its asphalt to expose the rich dark brown earth beneath it. Inside several deep trenches, men in blue shirts with trowels were squatting to some task. A sign on the chain-link said tours were available from 10:00 a.m. to 3:00 p.m. The gate was open.

When I stepped inside, a woman from Desert Archaeology, Inc. offered to show me around. She took me to a trench exposing a three-foot-wide, raised length of pinkish adobe bricks. It looked like a giant key fitted into a slot. "We think this might be the presidio wall," she said. Then she led me to the privy. "The lighter-colored earth tells us it was backfilled some time in the past, probably by a glass jar hunter. But here's one he missed." She showed me a pit house of the Hohokam, who settled here a thousand years before the first Europeans arrived to build their forts. In one of the storage holes, they had found a glass pickle jar. "We didn't expect to find it," she explained. "It was left by previous excavators in the 1950s." Inside the jar was a message to future archaeologists, asking for the preservation of the presidio.

In a few months workers backfilled the diggings and the parking lot returned. But then, in the early spring of 2006, the fence reappeared and I started seeing pallets of large adobe bricks. A sign credited the new construction to the Means Design and Building Corp. It looked as though workers were making at least some of the bricks on-site, using screened dirt shoveled from the presidio grounds and pressing it into wooden forms the size of sheet cakes. I was fascinated. Soon, the bricks became walls and scaffolding arrived as the walls grew thick and climbed to ten feet. At the northeast corner, the brown walls rose even higher, becoming a tower, what the Spanish called a *torreón*, complete with embrasures, or gun ports. Soon, the air would roar with cannon fire and the white and gold Spanish flag would unfurl over Tucson for the first time in more than two hundred years.

On Saturday, May 19, 2007, I parked downtown and walked to the celebration. A pair of presidial soldiers wearing black, wide-brimmed Texcuco hats, trimmed in scarlet ribbon, and dark blue *chupa* coats with

Spanish Presidio re-creation in downtown Tucson

scarlet collars, cuffs, and lapels stood guard at Washington Street where a crowd had gathered outside the *entrada*, or main gate. The Ronstadt-Ramirez Santa Cruz River Band sang ballads while people wearing shorts and T-shirts mingled with friars, presidial soldiers, including *soldados de cuera* (leather-coated soldiers), and bluecoat Catalonian volunteers. When a hooded, gray-robed Father Garcés suddenly appeared, I warned him, "Padre, stay away from Yuma!"

At 10:00 a.m., the local dignitaries began their speeches. "This was the beginning of Tucson," Mayor Bob Walkup said. "The presidio helps preserve our past and our future." Others chimed in with similar sound bites.

Then Gayle Hartmann, an anthropologist and former president of the Tucson Presidio Trust for Historic Preservation, read a letter from Emil Haury, the archaeologist who had left the message in the pickle jar. Next, Hector Soza, whose great-great-great-grandfather was a presidial soldier, read in Spanish Hugo O'Conor's August 20, 1775, letter to the King of Spain about his decision to move the presidio to Tucson. Finally, a

Franciscan brother offered a blessing, and the color guard of Catalonian volunteers in black tricorn hats and shouldering smoothbore muskets began to march toward the entrada to the beat of a redcoated drummer. "Gentlemen," someone shouted. "The gate is open. Fire the cannon!"

Instantly, an ear-ringing detonation reverberated down Church Street, and the Spanish flag appeared above the torreón.

I followed the crowd up the walkway and along the *torreón banco*, a bench of adobe bricks that covers and protects the original presidio tower foundation. Once through the gate, my eyes were drawn to a wide mural on the southern wall, a scene of presidio life with soldiers and citizens and livestock, the comandante's residence and presidio chapel to one side, the distant blue Santa Rita Mountains sweeping across its length above high painted walls. Seventy-eight-year-old Hector Soza stood with his long steel-bladed lance in front of a painted soldier who looked remarkably like him, right down to the five-four stature, dark hair and complexion, sharp nose and thin mustache. The two could have been the same person.

In the afternoon, I returned to the presidio celebration with Karen. The "comal and horno" (griddle and bread oven) spewed wood smoke and the smell of fresh tortillas. Hector Soza was still standing in front of his painting, recalling the famous "May Day Battle" of 1782 when six hundred Apaches attacked the presidio, at the time only a palisade of cottonwood poles.

With a voice full of language dating to the conquistadores, he said, "I was one of the twenty-five soldiers guarding the horses when the attack took place. I grabbed my weapons and saw that the enemies had split their forces. Half went to attack the Pima village, the other half trying to penetrate the fortress through the main gate. . . . The force attempting to cross the bridge was repulsed by the two cavalrymen already on the other side. . . . Meanwhile, Lieutenant Abate and twenty men joined in the defense of the presidio, repelling a swarm of Apache braves. Captain Allande's leg was hit by an arrow. He yelled at the sergeant to fire the cannon and when it finally fired its loud fierce noise, the Indians, who had never heard such a horrendous blast, retreated from the battle . . ."

It's a wonder we aren't all speaking Apache.

When he finished his story, Hector gave Karen a hug and introduced his wife, Mickie Soza, who was also wearing colonial-period clothing, her head covered with a black lace mantilla held with a comb. While Karen chatted with them about our local government and the city's transportation issues, including rumors of a plan to bury the interstate, I watched in amazement. The Sozas continued to care deeply about Tucson. There were many other fascinating events that day: Homer Thiel was on hand to interpret artifacts (like a Remington revolver) found at the restored 1866 Siqueiros-Jácome row house. The Catalonia Volunteers in yellow waistcoats gave a five-musket salute from a high rampart on the torreón. But standing next to the Sozas in that place, I felt honored to know two people so connected to a place that they could trace their family back generation after generation. I imagined Hector stepping into the same dusty tracks his ancestor Ensign José María Sosa left, and in his own way, defending and building Tucson's community.

I continue walking along Mission Lane to where the second component of the Tucson Origins Heritage Park will rise out of a bladed landscape. Soon, I encounter the paved pathway of the Santa Cruz River Park where Mission Lane once crossed the river and headed toward downtown. This is the southeast corner of the San Agustín site. Dozens of white PVC pipes stitch a dry canal, the city's latest efforts to deal with the underlying landfill before work begins here.

On this particular day, I'm fortunate to meet Michael Bodin and Jimmy Straley, who are monitoring the water lines to the landfill. "It burns it right up," says Michael with a wonderful western drawl. His accent—at first I think he introduces himself as "*Mark*ael"—and full gray beard make me believe he's a character from the 1880s. His sidekick Jimmy drives a golf cart, but I think the two should be on horseback. "There are only five people in the world who know how to do this," Michael explains when I ask him about the process of converting a landfill into real estate. "And I'm the fifth."

Soon, visitors to this part of the heritage park will walk the same route I just took from the mission gardens to the mission complex, following willow- and ash-bordered acequias through a re-created cienaga of cattails and reeds and shallow ponds. I can see it now: A wooden bridge leads me over the acequia and through the mesquite gates and whitewashed walls enclosing the San Agustín complex itself with its gleaming Convento and mission church. The place is alive with activity. Outbuildings model blacksmith shops, tallow and candle making.

Ocotillo pens hold turkeys and quail, while mesquite retaque corrals fence off live Spanish-era merino sheep, four-horned goats, Corriente cattle, and burros. Inside the Convento hangs a replica of the crucified image of the black Christ.

Along with the new mission gardens, cienaga, Convento, chapel, and granary, the twenty-acre park features an interpretive center with hands-on displays and an ancestral village with replicas of ramadas and pit houses and brush huts to represent an evolution of Native American life in the valley. The park also interprets life during the Mexican period (1821–56) with a reconstruction of Leopoldo Carillo's home near its historic location just south of the Convento.

It's encouraging to me to see how far the people of Tucson have come since the days when this place had value only as a dumping ground. Rivers, it seems, have always been places to hide our environmental transgressions. As if burying our waste or sending it downstream would make it someone else's problem. Trash, like our past, has a way of resurfacing. Better to recycle it in the first place. In my mind, Rio Nuevo's Tucson Origins Heritage Park is the greatest recycling project on the planet.

The Garden of Gethsemane and Sisters' Lane, 123 River Miles

Master recyclers Michael and Jimmy offer me a bottle of Gatorade and a ride in their golf cart to Congress Street, both of which I gladly accept. Here, just west of where the river path dips beneath the road, a eucalyptus sprays skyward in a giant inverted pyramid of white limbs and dark green leaves. This is Tucson's largest tree, the sign claims, a "fine example of Eucalyptus rostrada." The marker also says that the tree measures thirteen feet in circumference and has a diameter of four feet, and that "it is growing in a very favorable location, a region of deep rich soil and its roots are undoubtedly down to permanent ground water. It is young and vigorous and will be of great interest to see how large it will get. Let us guard it and protect it and give it a chance." The city is serious about this. In the 1930s, when Congress Street was extended west and widened, crews paved the road so that it curved around the tree.

Local history says Tucson's official tree was planted in 1910 by a young Menlo Park girl named Defina Bravo. Apparently, the girl rescued the sapling after a severe thunderstorm and flooding had uprooted it. The tree came to be called "Fina's Tree."

In 1910, twenty years after Sam Hughes's artificial headcut at nearby St. Mary's Road had turned the river into a deep ugly arroyo, every significant storm would have torn at the river's banks, toppling trees and people's homes. Brave Defina gets my vote as the Santa Cruz River's first restoration volunteer.

Across from the giant tree and tucked next to the Congress Street bridge is a wrought-iron and block-walled cloister, shaded by ash, olive, and mesquite trees, called the "Garden of Gethsemane." I walk through the open gate and I'm confronted with the smiling bust of a man who could be any homeless man living under any bridge. In fact, Felix Lucero, a Native American born in Colorado about 1895, *was* a homeless man living under *this* bridge.

A photograph of him in 1938 in the *Arizona Daily Star* shows him sitting in the riverbed next to his life-sized sculptures. The caption calls him an "itinerant artist." It's the eve of Good Friday and people have crowded the bridge and banks to watch Lucero work on a statue of the crucified Christ lying on the sand, the medium which Lucero originally chose for his creations.

Felix Lucero arrived in Tucson the year before the photo was taken and quietly began shaping damp sand in the Santa Cruz River into religious statues. Soon, he began drawing attention. When people asked what he was doing, he said, "I told God that I would work for him, that I would give twenty years of my life to him. Next year, the twenty years will be ended." Lucero must have felt some urgency in keeping his promise. For it was at the height of World War I, while huddled in a trench on a battlefield in France as German artillery decimated his battalion, that Lucero made his vow. If God spared him, he prayed, he would dedicate two decades of his life to poverty and faith and art.

After the war, Lucero left a trail of religious sculptures in twenty-seven countries, the Philippines, and California before coming to Arizona. His first work in the Santa Cruz River, the crucifixion image in the photograph, washed away in a flood. Then he started mixing cement into the sand, supporting himself and his work with odd jobs and the few coins—and an occasional sandwich—people tossed onto his spread blanket.

In 1946, after more flooding destroyed his work, Lucero moved to higher ground and began creating the figures we see today, a life-sized Christ on the Cross, Christ Entombed, the Holy Family, and his magnum opus, a scene reminiscent of Leonardo da Vinci and depicting the Last Supper, complete with Judas opposite the Nazarene.

Felix Lucero died of a brain hemorrhage in 1951 at the age of fifty-six, but not before finishing his work even though at the end he suffered horrible scars and missing fingers, the result of his shack catching fire the year before. After his death, the city of Tucson officially dedicated the Garden of Gethsemane to his memory.

Over the years, there were more episodes of flooding, and vandals nearly destroyed the statues on several occasions. In the 1970s, the city moved the park farther up the bank to accommodate the widening of the Congress Street bridge, restoring and fencing off the statues. They came to their present location in 1982 when the city realigned and channelized the river. Tucson Parks and Recreation built the present enclosure on June 13 of that year and honored the artist by naming it the Garden of Gethsemane Felix Lucero Park.

As they have for twenty-seven years, the Knights of Columbus maintain and care for the garden, where house sparrows chirp among oleander and pittosporum shrubs, and grackles squeal from a nearby tamarisk looming out of the river channel. Standing before Lucero's Last Supper, I wonder about the odd proportions of the white figures. Large noses, near-featureless faces. The hands of one disciple on Jesus' left appear to have fingers for thumbs.

Lucero was no Michelangelo. His work betrays the skill of a self-taught artist, but one thing is certain. He was dedicated to it. And now, nearly sixty years since his death, this place stands as a shrine to the creator as much as to his creativity.

Not bad for a man living in a cardboard box under a bridge and playing in the sand.

I stop at a mesquite-shaded bench along the river path for lunch—left-over fresh vegetables and grilled chicken from Melissa's Mars landing party last night. Twig-hopping verdins call excitedly and the male house finches have flushed red in this season of heat.

Afterward, I follow the path beneath St. Mary's Road, what early maps call "Sisters' Lane" for the seven sisters of St. Joseph Carondelet. The women arrived at Tucson on May 26, 1870, after leaving St. Louis, Missouri, a month earlier and traveling by train to San Francisco, by ship to San Diego, and then across the California and Arizona deserts in the back of a wagon.

The diary of Sister Monica Corrigan records the momentous occasion: "At about three miles from the town we were met by the procession which was headed by four priests on horseback; but as we came in sight, they dismounted, and ran rather than walked, to meet us; the crowd, in the meantime, discharging firearms. Before we reached the city, their number had increased to about three thousand; some discharging firearms, others bearing lighted torches; all walking in order, and heads uncovered. The city was illuminated—fireworks in full play. Balls of combustible matter were thrown in the streets through which we passed; at each explosion, Sister Euphrasia made the sign of the cross."

The seven women—Monica Corrigan, Euphrasia Suchet, Emerentia Bonnefoy, Hyacinth Blanc, Martha Peters, Ambrosia Arnichaud, and Maxime Croisat—had overcome a month of hardship to fulfill the obligation of their faith in whatever capacity the frontier town needed: nursing the sick, caring for orphans and the poor, or teaching trades to prisoners and the disabled.

The religious order dates to 1650 when, in LePuy, France, a Jesuit contemporary of Father Kino named Jean-Pierre Médaille began organizing small secret clusters of three nuns each to minister to their neighbors. By 1836, they had sailed to the New World and established their first North American house in Carondolet, just outside St. Louis, Missouri.

In Tucson, the seven sisters immediately opened Tucson's first school, which soon grew and divided into a secondary school, St. Joseph's Academy, and an elementary school, St. Augustine's. They learned both Spanish and O'odham for their Indian school at San Xavier. Within years the sisters had started a novitiate for local girls who expressed interest in joining the society. By the end of their first decade, the Sisters of St. Joseph were a common sight in Tucson, their black habits and veils billowing after them as they hurried along its dusty streets.

When the railroad arrived in 1880, it came with a daily toll of injured men from the train and track gangs, machine shops and fuel depots. As suddenly as Tucson entered this new age of speedy transportation, the community needed a hospital. The burden fell to Father John Baptist Salpointe, Arizona's first bishop after the Catholic Church granted the state the "pre-diocese" jurisdiction status of a vicariate apostolic in 1868. Bishop Salpointe had purchased land a mile and a half west of town next to the sisters' novitiate for girls and had nearly completed a trade school for the local Native American youth. But the school could wait, the bishop regretfully determined, and turned the building over to the sisters to use as a hospital. The story goes that the O'odham boys were deeply disappointed, but nevertheless helped finish the hospital when they saw the women struggling with some heavy rocks.

St. Mary's Hospital, the oldest hospital in Arizona, was dedicated on April 24, 1880, and admitted its first eleven patients on May 1.

Those same early maps also show Sisters' Lane (St. Mary's Road) as the approximate location of Sam Hughes's infamous headcut. At 3:00 p.m., I stand under an enormous tamarisk just north of the road, scanning the Santa Cruz River for any vestiges of the ditch. Ironically, the river is probably much deeper now than Hughes's ditch ever was.

Like many of his American entrepreneur contemporaries, James Lee and Solomon Warner among them, Samuel Hughes rode the high tide to Tucson following the Gadsden Purchase. Successful in California's gold rush but suffering from tuberculosis, the Welshman arrived at Tucson in 1858 and found the drier climate more to his liking. (Some say his stagecoach driver actually dumped him here to die.) He immediately opened a butcher shop, and soon enlarged it into a general store. During the confederate occupation, also like Lee and Warner, Hughes refused to live under a Confederate flag and retreated to California. He returned in 1862, marching with Colonel Joseph R. West and the boys of the California Column.

On March 27, 1863, Sam Hughes married a beautiful Mexican girl named Atanacia de Santa Cruz in the mission church at San Xavier. Hughes was thirty-three; Atanacia was twelve. (Some records show them marrying the year before, on May 27, 1862.) The two had probably met a few years earlier at the home of his soon-to-be business partner Hiram Stevens, where Atanacia lived with her sister Petra after the death of their parents. (Hiram had married Petra a few years earlier when she was sixteen.) At first, Atanacia put off Sam's advances, claiming she didn't know how to cook or wash. Sam assured her, "You'll never have to wash."

Atanacia proceeded to bear him fifteen children.

By the mid-1870s, Hughes was the largest single landowner in Tucson, particularly of agricultural land. Hughes wasn't a farmer, however, but a businessman, a land speculator, looking for quick money. Instead of learning from the local Mexican or Tohono O'odham farmers and utilizing their more conservative practices, Hughes arrogantly went his own way, like excavating a channel to the subsurface aquifer across the Santa Cruz River, intending to irrigate fifteen thousand acres with the resulting flow. In the end, the farmers and the environment suffered the most. Hughes, on the other hand, along with his Anglo newcomers, remained untouched, even profiting from the devastation as more and more of the original farmers couldn't afford the resulting but necessary schemes to water their crops.

And there were the farms directly lost as Hughes's folly continued year after year to claw its way south, giving us the ever-widening channel of blinding sand we see today.

I think about Sam Hughes a lot these days, especially when I stop for a red light at Oracle Road and Ft. Lowell. His grave in Evergreen Cemetery is marked by the largest stone just to the west of the intersection. I'm now advocating for a change in names in his honor: the West Branch should be returned to its former name and original glory as the Santa Cruz River. The current "Santa Cruz River" designation should be changed to "Hughes's Ditch." Not to be confused with "Huge Ditch."

This part of the Santa Cruz River Park on the west side between St. Mary's Road and Speedway is my favorite, and I've walked it or ridden my bike along it many times. Today, the jojoba shrubs (female plants) are heavy with acornlike nuts. Mature mesquite, desert willow, and blue paloverde trees mottle the path with shade, while thick clumps of globemallow spray colors ranging from peach to magenta. Among the grassy pocket parks with their benches and tables and barbecues are gateway features in brilliant stone and tile mosaics representing the local river neighborhoods: "Anita," "El Rio," "Hollywood," and "Pascua." From somewhere along Riverside Road, church bells ring out "America the Beautiful."

Soon, I pass beneath Speedway Boulevard and walk along a line of ancient tamarisk trees bordering the Arizona School for the Deaf and the Blind, founded upon our statehood in 1912 as the Arizona State School for the Deaf. I imagine the trees were planted with the school. Some of their trunks are huge, maybe six feet in diameter. Lesser goldfinches whistle plaintively among the branches—it must be twenty degrees cooler in their shade.

At 3:40 in the afternoon I approach Grant Road where Karen will meet me. I'm spent, my legs tired and sore. The sun is still hot and lays

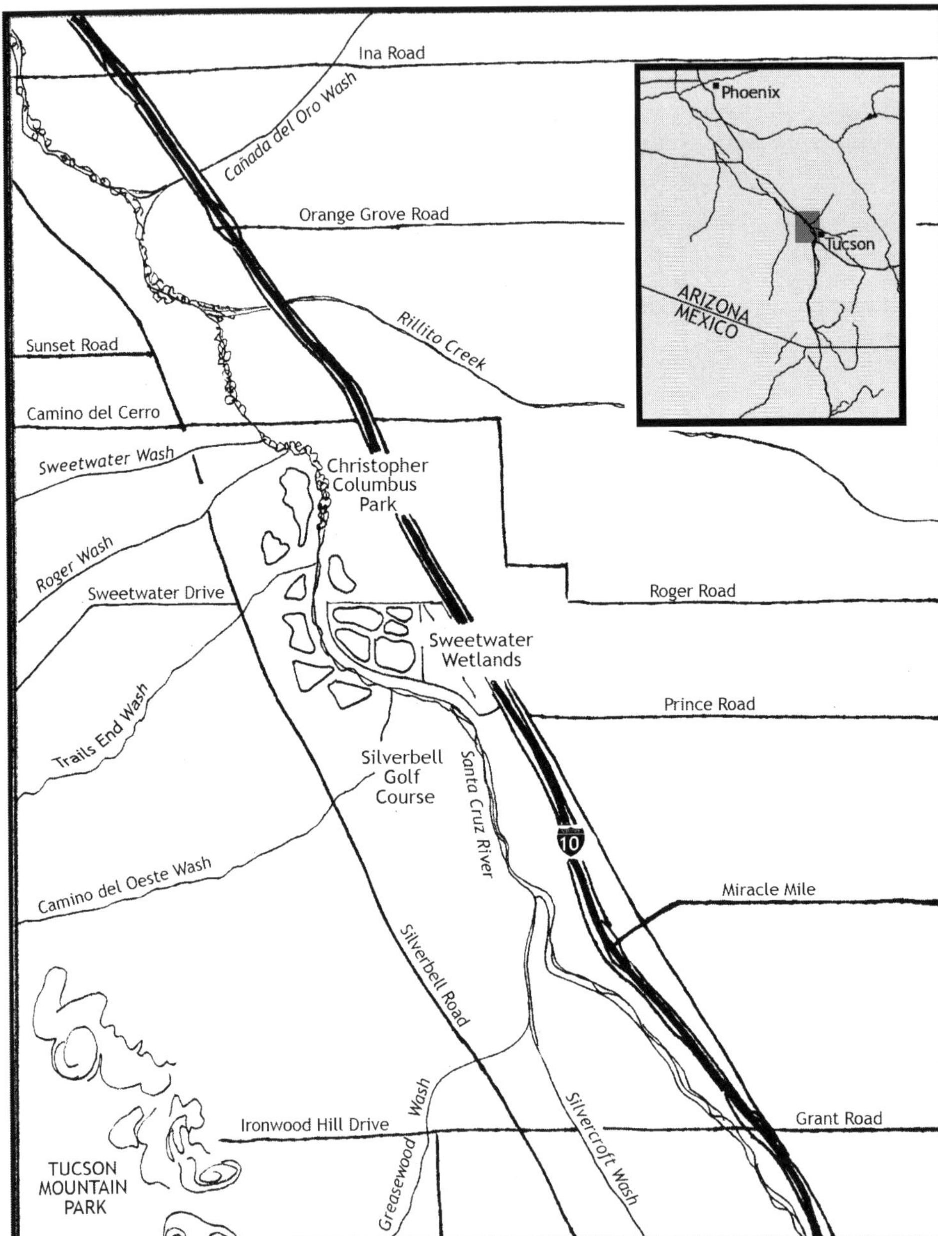

The Tucson Reach (Sweetwater River)

itself firmly against my already reddened skin where the young mesquite trees offer only thin comfort. There are, however, signs of a much older bosque that once crowded the riverbank here in the occasional giant gray stumps that lie in the grass like stone altars.

Sweetwater River: June, 126 River Miles

In early June, on what promises to be the hottest day of the year so far (104°F), I arrive at Grant Road and the Santa Cruz River with great intentions of hiking eight miles or so to the Ina Road bridge. It's 6:00 a.m. I park my 1979 Datsun 210 in the meager shade of a paloverde tree, having made arrangements with my daughter Jessica to pick me up in five hours.

The Santa Cruz River Park pathway has ended, but the railing-topped, gunite-stabilized bank provides a nice walking path on both sides of the river, so I choose the west bank, stepping high above a desert broom- and burrobush-choked ditch of sand. The channel seems to amplify the sounds of the freeway and a distant train whistle but also the clear ringing *peek*ing of an Abert's towhee.

Where the river channel shifts to the west, a red-tailed hawk, perched atop a power pole beside some river-bottom pasture and cornfields, stretches its body and takes its first morning flight. My path now drops away on both sides, and I walk along a narrow section of berm that smooths out and reshapes the river's formerly wide personality swings, which once cut deeply into its western flank. What remains on my left is a wide half-moon of gouged land, cut off from the channel on my right and open to brush, sprinting jackrabbits, and the makeshift shelters of the homeless.

After an hour, I've crossed the arching ironwork bridge over Silvercroft Wash, whose source begins in Starr Pass behind my home. From there, the arroyo cuts through the desert next to where my girls went to elementary school, slides west of Tumamoc Hill and along St. Mary's Road, and rushes past the grocery store where I shop, soon picking up Greasewood and Painted Hills washes before joining the Santa Cruz River in what must occasionally be a mad torrent of runoff.

Here, the water hazards and fairways of Silverbell Golf Course give Gambel's quail and golf balls a place to mingle beneath tall Aleppo pine and eucalyptus trees. The concrete-lined Camino del Oeste Wash splits the greens evenly after coursing out of its source in Gates Pass and the backyard of the abandoned "Stone House" on my favorite trail named for my favorite social scientist, David Yetman.

The Yetman Trail winds through an area of rugged desert where Jessica has set three of her infrared "trip" cameras for her mountain lion/bobcat project, and she's probably off there today downloading their digital images. Along with her Sky Island Alliance jaguar work, Jessica has teamed with Lisa Haynes, a wildlife biologist and University of Arizona researcher studying southern Arizona's big cats. The project is exploring several major concerns, such as the current population status of mountain lions and bobcats in the Tucson Mountains, how these animals are moving through the region, and the effects urban development is having on them. "The population seems fairly healthy in the Rincon, Catalina and Santa Rita mountains," says Lisa Haynes, "but less so in the Tucson Mountains. We don't know if the lions can get in and out of the Tucson Mountains, or if the population is isolated and possibly inbred because the range is surrounded by housing, roads, railroad tracks, and the Central Arizona Project canal."

Jessica spends many hours each week hiking along these rocky arroyos like the Camino del Oeste and Sweetwater washes, monitoring dozens of cameras to learn the answer. My feline-fixated Jane Goodall. It's amusing for me to think of my little girl as a "wildlife biologist," collecting droppings and hair and animal parts for genetic tests. It's true that she chased the boys around her grade school with monstrous beetles she would find and catch. That she went everywhere with a white rat named "Dolly," tucked beneath her blonde hair as the rodent rode on her shoulder. (Oh, the screams she elicited from the women in her grandmother's church group when, after chatting with her, they suddenly noticed that pink, hairless tail slip out from under her curls!)

All the same, I still remember her squeamishness-bordering-on-hysteria when it came to her encounters with all the bloody, vomiting, defecating, stinking pets her sisters kept—and the minimally gutted, half-devoured, partly decapitated, and mostly unidentifiable victims their pets dragged into the house.

One such pet was "Pern," whose messiness was an order of magnitude above all others. We carried Pern home in a small, hole-punched cardboard box pet stores use for packaging their mice and birds. Kasondra and Melissa, who were both in elementary school at the time, decided on the name because they had become enamored with a series of books by Anne McCaffrey about a planet inhabited by dragons. The author called the world "Pern," which as I remember was an acronym for something. So, my daughters named our tiny green iguana for a mythical dragon-world populated by a few people who had learned to build a society around the beasts. The girls should have given the name to our household instead.

I bought the iguana a cute little leash. Karen, always the smart one, returned to the pet store and bought a book on iguanas. Then she read it. When she finished, she said, "Just another small thing you've given me that grows up!" Karen never minced words. "And this one is worse than a baby. At least babies wear diapers."

Green iguanas, we learned very quickly, have special needs. Because they come from rainforest climates in Central and South America, the lizards pale to our rock-tough, desert lizards. Iguanas require a cooler, more humid environment than our hot, desiccating landscape can offer. They want to be indoors in climate-controlled rooms. They like to be misted.

And this is just the beginning. Because they now live indoors, you must provide belly heat and full-spectrum sunlight for them. This enables your new iguana to properly digest all the fresh bananas and mangoes, spinach leaves and squash blossoms you feed her on a daily basis, when you're not misting her majesty as she basks on her heat rock under her sunlamp.

Pern adjusted well to our home. The girls created a place for her in a ten-gallon tank with a basking rock and tree branches to climb and a "pond" to bathe in. Pern loved water, and always took bath time as an opportunity to drink deeply and then relieve herself. (Melissa and Kasondra once tried to put Pern in the bathtub, but Jessica threw a fit, screaming, "I'm not soaking in a tub that Pern pooped in!") They took her for walks on her leash or rode around on their bikes with her gripping tightly to a shoulder. I still have photographs from this time when she was small: Pern with her oversized leash on the porch fence; Pern perched on the head of smiling Kasondra. But soon she no longer fit her leash. Then she outgrew the ten-gallon tank, its thirty-gallon replacement, and had begun to look uncomfortable in the fifty. From the beginning, I knew she had designs on the living room, the largest room in our house, and that she would insist on some modifications.

Karen found the birdcage, a refrigerator-sized, wrought-iron monstrosity that she felt Pern must have to be comfortable living with us. I believe a giant parrot or possibly a condor had been the cage's former occupant. Three hundred dollars later, with some added shelves, hot rocks, and sunlamps, Pern became furniture in our living room. The only furniture. Since the room wasn't large enough for a couch *and* Pern, the couch had to go. We had no place for anyone to sit in our living room, but we did have something interesting just above eye level to look at while you were standing there. Something that always looked back and down on you, usually with smug disdain.

Pern became the center of attention in our family. From her high perch, she examined our comings and goings. She watched television with us. She played games with us. When we pulled out our dining table and set chairs around it, she ate meals with us. This is where Jessica would begin to hyperventilate, and not just about the sounds and smells. "Why do I always get the sneeze seat?" she would complain to her sisters, who always left her the chair closest to the cage.

Iguanas have a particular way of removing excess salt from their bodies. Special structures in their nasal cavities collect the salt, which the animal then combines with liquid and forcefully ejects. The behavior doubles as an annoyance mechanism, intended to alarm those who come too close or, as in Jessica's case, to thoroughly disgust them.

It worked like this: Jessica would sit at the table in her assigned chair. Pern would maneuver on her shelf to line up Jessica in her sights. Just as my daughter began forking food into her mouth or drinking from a glass, Pern would execute a short nasal burst, freezing Jessica in mid-gulp.

"Pern!" Jessica would shout. "That's so gross!" To which Pern would only respond with a satisfied grin.

Today, Jessica has no qualms about bagging scat or scooping up sun-ripened roadkill. She still speaks fondly of her first javelina necropsy, discussing vivid details matter-of-factly over greasy plates of Lucky Wishbone chicken or Silver Saddle steaks.

From the eighteenth hole at Silverbell Golf Course, I look across the Santa Cruz River into the impressive cottonwoods of one of the hottest of bird-watching hotspots in southern Arizona.

Sweetwater Wetlands Park gets its name from Sweetwater Wash, which drains a major watershed beneath Wasson Peak in the Tucson Mountains before entering the river a bit farther downstream. The wash traces critical desert riparian habitat and is an important wildlife corridor that connects Tucson Mountain Park to the river.

Recently, Pima County acquired a key segment of this corridor for conservation. Sweetwater Preserve, 695 acres of saguaro-clad foothills, is one of the largest remaining undeveloped private tracts of land in the Tucson Mountains, home to gray foxes, tiger rattlesnakes, leaf-nosed bats, and even some of Jessica's mountain lions.

"Sweetwater" probably refers to the actual taste of the water that once bubbled to the surface near the Santa Cruz, so I suppose it's only ironic coincidence that an artificial wetlands created by Tucson's wastewater would bear the same name. A committee of environmental and education experts directed the construction of the wetlands in 1996, opening it to the public in 1998. Relying on backwash from the nearby Roger Road Wastewater Treatment Plant, Sweetwater moves effluent from settling basins, through a series of polishing ponds, and then finally into the recharge basins, where it percolates into the aquifer. Reclaimed water needed for Tucson's parks, golf courses, and school playgrounds is then pumped from the aquifer and chlorinated before delivery.

It's the polishing ponds that attract the wildlife and in turn the wildlife watchers. While aquatic plants like bulrushes and cattails create an environment for microbes to convert nitrogen into nutrients, they also provide the basis for a wetland ecosystem of insects, amphibians, reptiles, and birds.

Yesterday, I walked the paths among the giant four-wing saltbush and towering willow and cottonwoods ringing the wetlands. I was on a mission for one particular bird. In the tall bulrushes and cattails, red-winged blackbirds gurgled liquid songs while fat desert spiny lizards blundered through the cornflake litter of saltbush. At the Gazebo Pond, dragonflies in orange and blue neon rattled in figure eights above red-eared sliders paddling through water like pea soup. Everywhere was the smell of life's pungent by-products.

I came hoping to see a gray hawk, the short-winged migrant of riparian woodlands like those along the San Pedro River, and my bird of hope. This wasn't a fantasy of my own concocting. A year ago last May, bird-watchers Rick Taylor and Clive Green spotted an immature gray hawk at these ponds.

This past spring, Sweetwater Wetlands has been receiving a lot of attention in the birder community. On May 5, 2008, weekend bird-guide Jerry Bock, on his usual rounds at Sweetwater, was stunned to see a flyover swallow-tailed kite. Only once before, more than twenty-eight years ago, has this graceful avian stealth-fighter been recorded in Arizona.

Even more extraordinary, only last week Jean Barchman reported a juvenile female elegant trogon at the nearby Roger Road plant. The following day, the bird showed up at Sweetwater with an adult male. Elegant trogons, shining emerald and geranium birds of the tropics, usually haunt the oak-draped canyons of southeast Arizona's sky islands. These are the first ever seen at Sweetwater and maybe in the Tucson valley.

Almost exactly seven years ago I saw my first elegant trogon, and I remember the day with great fondness. The girls and I had rented our favorite cabin in Madera Canyon, Kubo #4, with the intention of passing the first week of summer vacation relaxing in the oak and sycamore shade. Outside the cabin next to ours, three researchers from Northern Arizona University had stretched a white sheet over the ground and set up an ultraviolet light at its center to collect insects. I talked to one man who stood over a wet seep, netting Arizona sisters butterflies and twin-tailed swallowtails. When I asked him if he'd seen any interesting birds, he said, "Yesterday I ran into a pair of trogons at the first stream crossing on the Supertrail."

It was just what I wanted to hear. I stuffed binoculars, camera, and water into my knapsack and grabbed Jessica, Kasondra, and Melissa. I knew the place and convinced the girls that it wasn't far up the trail to Mount Wrightson. "An elegant trogon!" I said. "I've never seen one—you've never seen one!" They were less than thrilled but agreed to humor their dad.

As we approached the stream crossing, I slowed our pace. Several large sycamore trees spread their smooth white limbs over us, and I searched among the bright leaves. Somewhere a dark hole had provided the birds a nesting cavity. Then, Kasondra looked up and said, "Oh, there it is," and pointed to a bird glowing in red and green neon on a juniper limb. The girls looked at the bird, looked at each other and then at me, and said, "Can we go back now?"

"Do you know how many people travel all across the country hoping just to glimpse this bird?" I said, but they kept silent, unimpressed. We walked back down the trail with the male trogon calling in a low, echoing voice, like a weak battery trying to crank a stubborn engine: *koa, koa, koa, koa, koa.*

The following March we returned to the cabin and to rumors of a male elegant trogon that had overwintered in the canyon. "Only yesterday he was feeding on berries in the ivy behind Kubo Four," Cora told us when we arrived late Friday night to check in. The next morning, I stood with my coffee at the large picture window looking out over the creek where ivy had completely entangled the trunks of three sycamore trees. Hard, dark purple fruit hung in clusters. As I sipped my coffee, Karen appeared at my side and said, "Are you going to go hiking for the trogon?" "Yes, *we* are," I said, as Kasondra and Melissa joined us, smiles crossing their faces. They were up to something. "Why?" the girls asked. "It's right there in the tree."

I couldn't believe it. A flash of bright geranium darted from the vines and alighted on a dead snag in full view not fifteen feet from the window. We hadn't even left the cabin.

The evidence is only anecdotal, but it seems to me that something historic is happening. A hundred years ago, no one reported trogons in Arizona, and now we have two species, elegants and eared quetzals, not just visiting but actually nesting here. And other tropicals like flame-colored tanagers, too. Evidence of global warming? Maybe. Just this year a short-tailed hawk appeared over Tucson, flying over the busy intersections near Wilmot Road of all places. Short-tails, forest-hunting raptors from Mexico and Central and South America, were first identified in southeast Arizona in the mid-1980s. And last month a tufted flycatcher, only the fourth recorded in the United States and the second in Arizona, attracted scores of bird-watchers to the Chiricahua Mountains. Add the birds to the other animal immigrants from the south—Mexican opossums, coatis, javelina, jaguars—and it really makes you wonder what's going on.

I didn't find any gray hawks yesterday at Sweetwater, or even hear one call. But I did encounter a great blue heron, the bird huge and holy in its dark feathered robe. I decided to believe that one day soon I would see the hawks return here, and that, for now, this sign of faith was enough.

At Christopher Columbus Park, I smell the full-bodied stink of effluent. This, however, isn't a Sweetwater smell but something much richer and darker, like comparing a Raging Sage French roast to Folgers in the can. This smell is caffeinated.

I cross the river bottom heading for a knotted rope of trees rising from the far bank and unwinding downstream as far as I can see, following the rushing sound of water and stepping over the sandy dimples

Santa Cruz River north of Roger Road

of rabbit tracks and tail-drags of lizards. Tiny burs penetrate my socks and needle their way inside my shoes.

When I reach the shade of the largest willow, I drop my pack and remove my camera. Water gushes from a huge culvert and falls into the river, launching into a terrific upwelling pool of white noise and foam. Here, at the "headwaters" of the lower Santa Cruz River, the outfall from the Roger Road wastewater facility transforms the Santa Cruz from a hot, dry, bank-stabilized ditch into a river gallery of willows and cottonwoods, not unlike, I think, the Santa Cruz of a hundred and fifty years ago.

The Roger Road facility is Tucson's oldest sewage treatment operation, constructed in 1951 and now slated for decommission as our newer plant five miles downstream at Ina Road recently completed its expansion. When Pima County officials first began considering the closure, pumping the plant's sewage through a conveyance line to the Ina Road plant, it caused a stir in the environmental community, especially among bird-watchers. Shutting off the flow of effluent into the river here "would turn it into a desert," said Phil Rosen in a January 2, 2006, *Arizona Daily Star* article. "They are just thinking like engineers," he went on, "without considering carefully the ecological aspects."

The effluent discharge at Roger Road makes this part of the Santa Cruz River the third largest perennial stream in eastern Pima County, a human-made corridor of teeming wilderness right at our doorstep. This would all go, along with Sweetwater Wetlands.

Fortunately, the county is now recommending a different option, one that isn't the least expensive but takes into account the ecology of the river. It's called a "water campus," a new reclamation facility constructed next to the aging and obsolete Roger Road plant and interconnected with the Ina Road plant. The water campus should be completed by 2015, and as part of the Tres Ríos del Norte project will serve to support an eighteen-mile "greenbelt" along the river from Sweetwater Wetlands to Sanders Road in Marana, three-quarters of the way to the Simpson restoration site where my daughters and I began working to save the Santa Cruz River almost six years ago.

Our sweetwater river. I've heard stories that the Santa Cruz River may rise suddenly, as if the tide in some nearby estuary were changing and sending a tidal bore upriver. It can happen quickly enough to catch river-walkers off guard, especially early in the morning on a weekday as hundreds of thousands of people getting ready for the workday begin flushing their toilets.

The gunite reinforcement on the west bank vanishes, and I hike at river level across a broad alluvium where two large washes enter from the west, probably Roger and Sweetwater washes. This could have been a marsh at one time. Today, a humming motor siphons groundwater from some deep aquifer—I wonder if the water is still sweet.

I duck into the cool-shaded willows and a Cooper's hawk darts by, chased by a smaller bird. The birds seem reckless with their voices as well. Bell's vireos, Lucy's and yellow warblers sing in various harmonies of *cheedle, cheedle-chee* and *wheedle, wheedle-whee* together with the occasional bright *whitchy, whitchy-whitch* of a common yellowthroat.

In 2000, a yearlong federal study found 112 species of birds on the river between Sweetwater Drive and Ina Road, while a later study found a rare yellow-billed cuckoo, a long-tailed woodland relative of our common roadrunner. In 2004, a LeConte's sparrow, a tiny, secretive grassland sparrow with brightly patterned black streaks over buff-yellow, was found just south of Ina Road. It is only the second confirmed sighting of the bird in Arizona. The updated (10/03/06) bird checklist I picked up at Sweetwater on Monday lists 230 species for the wetlands, which will now have to include elegant trogons.

After passing beneath the Camino del Cerro bridge, I enter what I consider to be the finest part of the river, a historic Santa Cruz. Sunflowers bloom over a wide floodplain erupting with dense thickets of mesquite trees. Along the dark stream, a black phoebe perches on a bare snag watching for passing insects not already swooped upon by the patrolling rough-winged swallows. Lesser goldfinches chortle and squeal from the willows, while more Lucy's warblers send their sweet, cheerful songs from the cottonwoods.

At 10:00 a.m. the heat is on. I rest in the shade of some of the largest willows I've seen so far. Behind me, a pole-sitting red-tailed hawk screams at my presence. Suddenly, an owl flushes from a willow and heads downstream. At first I think it's a great horned owl by the size of it, but then I see that its color is all wrong, a much lighter brown, almost tawny like a barn owl's plumage. I drop my pack and pull out my binoculars, finally relocating the bird farther downstream. When it turns its oval face away from me I see two tiny dark tuffs of feathers above its yellow eyes—a short-eared owl! Amazing! Short-eared owls normally spend the summer in northern Canada and Alaska, floating across open prairie and tundra, sometimes in daylight, searching for mice and other rodents. They range nearly worldwide, in North America dropping south into the middle of the United States in winter, but are considered fairly rare in Arizona, especially for June.

Trogons from the southern tropics and owls from the northern tundra—this reclaimed part of the Santa Cruz River west of Tucson is linking wildlife across the entire Northern Hemisphere.

When my cell phone rings I think it's Jessica wanting to know my status, but Kasondra is calling from Phoenix to tell me about the school district she'll be working for. Three weeks ago she married Chris, an air force first lieutenant she'd been dating for a couple of years, and the two moved into a cute rental in an old Phoenix neighborhood. Over the last month, her life went like this: graduate, get married, leave home, start teaching. Today, she's calling from a dorm room at Arizona State University where she's in training to use her public health degree in a low-achieving elementary school as part of the Teach for America program.

Santa Cruz River at Camino del Cerro

Willow gallery on the Santa Cruz west of Tucson

Like her sisters, Kasondra has also been following a trajectory in her life. If Melissa is my Carl Sagan and Jessica my Jane Goodall, then Kasondra is my Albert Schweitzer. Her passion lies with her concern for humanity, especially children who are disadvantaged and poor and largely ignored. In junior high and high school she volunteered at the International Wildlife Museum, running educational camps for kids. In college, she volunteered at the Ronald McDonald House, caring for seriously ill children and their families. She also became a Big Sister, and interned at Arts for All, a nonprofit supported by AmeriCorps that merges human services with the arts, bringing the arts to children with special needs. From teaching children about the world around them, whether it's the wildlife outside their doors or the physiology inside their own bodies, her path has been clear. I imagine that when her service with Teach for America is complete, she'll turn her eyes once again on Africa and the Peace Corps, and drag along her husband.

Husband. I'm still having trouble wrapping my mind around the word as associated with my daughter. It's not like I didn't think it could happen; I've seen it coming for some time. Since last summer, in fact, when Kasondra brought Chris along on *our family* trip to the beach.

Like the famous barn swallows of the famous mission, my daughters and I have been returning to San Juan Capistrano every year—only we arrive from the desert at the beginning of July and stay with relatives. It's become a seasonal migration, escaping the heat of our Tucson home to visit southern California and sink our feet into sand that, for a change, is wet and foaming. The beach is essential for reasons that include sun and the annual grunion run . . . but I can hear Kasondra say that I left out "boys."

Now, I like to remind all of my daughters that if they insist on husbands, I'm all for arranged marriages. I would like to spare you the grief of courtship and romance, I tell them. The reason romance is so attractive is that it invites risks, which heighten the experience and make you feel alive. So why seek romance when you can stand in three feet of dark ocean with millions of swarming, shin-bumping grunion? Grunion are the sexiest of fish, and I'm still trying to convince my daughters that a grunion run with Dad is better than any beach with boys.

We learned about the fish five summers ago. I still have the article, taped inside my journal. A picture shows squirming grunion: seven males horseshoed around a spawning female poking out of the sand, one dark eye searching the camera lens.

The author, Paul M. Young, writes about the history of the "fish out of water," and how some people doubt their existence because searching for them is like hunting for the mythical snipe (or perhaps the elusive Santa Cruz River sand trout). He gives dates and times for the runs, when high tides are conducive to launching the fish up the beach for egg-laying and fertilizing. He also offers recipes for those with culinary inclinations. "Keep in mind," he says about eating the fish, "that in grunion circles, it's not a great way to end a date."

But what about ending the date on the night of a grunion run with a bended-knee, tuxedoed marriage proposal? I should have known that grunion runs are rarely innocuous when it comes to romance.

The girls and I had always chosen Salt Creek Beach, only twenty minutes from San Juan up the Pacific Coast Highway. On this particular evening, as usual, we followed a long stairway that dropped us through patches of bank-hugging ice plant and brushy chaparral before finally spilling us onto a narrow beach.

The ocean's atomized brine filled our pores. The setting sun broke into a thousand strobes on its dark surface, which retreated from the shore to lay its enormous body against a flat and unbroken sky. I'd forgotten how different the desert is from the sea, and how much the two are the same. The desert, too, diminishes you in its incomprehensible vastness.

Darkness came as we waited on the sand, backed against a jumble of rocks above the high tide line. White scarves of fog flagged landward. Eerie blue phosphorescence pulsed along the crest of each wave as it curled toward shore. In the darkness, we settled back to watch, pulling blankets over our shoulders and huddling together for warmth. Occasionally, couples walked by us, and I could tell by their conversations that they had no idea we crouched nearby.

Then it happened. Suddenly, Kasondra said, "Oh, look!" And there, only twenty yards away, stood a couple dressed in formal wear, he in a tuxedo and she in a pink chiffon evening gown and bare feet. Unaware of us, they danced slowly, arms around each other, her head on his shoulder, while music played from a portable stereo.

After a few minutes, Melissa spoke: "Look! He's down on his knees!" My daughters became ecstatic. "He's holding out something," Melissa continued, "a box—it's a ring! Oh my God! He's proposing to her!" Now the girls were all atwitter. I started to say something and Jessica jabbed me in the ribs and covered my mouth. I pulled away from her and shouted, "No! Don't do it! Run for your life!" But the words came out garbled as more hands pummeled me.

Then he was back on his feet and holding onto the girl again. She held the box and had begun to cry.

"She said 'Yes'! She said 'Yes'!" Melissa repeated.

"They're doomed," I said.

After the couple left, we seemed to be the only people remaining on the beach. The girls dashed madly to the surf, daring the breakers to catch them. Jessica rolled her jeans into sodden masses at her knees and raced along the surf line, long blond curls flying behind her. Melissa practiced karate, her movements like a dance set to the music of crashing waves. Kasondra, her long legs kicking against the sand, ran up and down the beach, in and out of the water, twirling and falling and swinging her arms, dragging around ropes of seaweed in a wild frenzy that could be matched only by the grunion crowding offshore.

Then, our flashlights started reflecting quick slips of silver shooting up the beach with each wave front. Melissa yelled and darted into the waves to scoop up one in her hands, but as the water retreated, the

Kasondra Lamberton and a grunion

blue-green knife of a fish wriggled out of her fingers and escaped over the sand as she grappled for it with both hands, shouting, "Hey, hey, hey!"

The grunion were massing. Soon, hundreds were streaking like blue meteorites up the flat wet sand, their images pulsing in our lights. Where we stood in the water, they thumped against our legs in their ecstasy to ride the next wave and spawn.

I've always admired grunion. No messing with long courtships, no emotional hang-ups with relationships and their multiple contusions and bruises. Just one frenetic night of activity to propagate the species and it's back into the wide ocean.

Romantic love, on the other hand, is so treacherous. It invents complete fabrications in your mind about yourself and the one you love. Nothing else matters. It has caused the downfall of civilizations. Love is a potent poison, a venomous snake . . . and it is the most wonderful thing that can happen to you.

I'm the one who was doomed. How many years had we come to that beach for the summer grunion? My gift of grunion, I realize now, pales beside the one offered in a small velvet box held by a lover's hand.

That night, their young bodies blurring against the white sand, my daughters were also massing, preparing for lives beyond mine when, according to nature's incomprehensible vastness, some near-future moon would raise tides already surging within them, drawing them away from the dry desert river where they were raised to other shores to begin families of their own.

By 10:30 a.m., I make it to the Rillito River confluence where the Santa Cruz spreads out and meanders over a flat, sandy plain. The Rillito, a dry wash deeply incised at this point, must be half a mile wide. A slow backwater slips into the drainage and mirrors its high, gunite-stabilized north bank. Here, I understand, will be a future river-walk connecting the Santa Cruz River Park with the Rillito River Park.

I move on, but the walking becomes a slog over hot sand that tugs at my legs as I lose traction. Even worse, the shade is gone and I'm wishing I had worn my hat. I feel my sunscreen growing ineffective.

I've had enough.

Knowing Jessica must be on her way, I bail out just past where the Cañada del Oro Wash shoves in from the east with a load of Catalina Mountain sand. I'm within sight of the Ina Road bridge, but instead I cut through an uneven terrain of dry grasses and creosote and take a dirt track to Silverbell Road.

"Almost made it," Jessica says when she picks me up, handing me a bottle of Aquafina from the cab of her truck.

The water is hot.

Yuma Wash, 133 River Miles

Today, the day will be twenty-seven seconds longer and the sun will set at seven thirty. I drive to the Santa Cruz River at Ina Road bridge and park on the northwest side to watch this evening's display of lesser nighthawks as they lift themselves from the surrounding fields to screen the hot air of mosquito plankton. The feeding has already begun when I arrive.

From under the bridge, water cascades over concrete, the rushing flow from the Ina Road Water Pollution Control Facility joining the Roger Road plant's output in a wide meander. The rich smell hangs in the air like the winged insects, which glitter in the angling sunlight.

Santa Cruz River at the Ina Road bridge

Nighthawks congregate above several large willows, dipping and swinging languidly against a darkening sky. There are too many to count.

I walk along the railing atop the gunite-stabilized riverbank while killdeer cry from the waterline and a pair of agitated black-necked stilts complain incessantly with keening voices. They circle overhead, their slender black wings pumping and their obscenely long red legs dangling behind, wasplike. Momentarily, a second pair starts up, and then a third. All the stilts come in pairs. When I finally think I spot a loner, the bird turns out to be an American avocet, another long-legged shorebird in a striking black-and-white tuxedo.

I grow self-conscious: The entire neighborhood is yelling at me.

Less than a mile downstream from the bridge, a ridge of outwash on my left fans into the river-stabilizing berm. The sediments are debris from the Tucson Mountains carried here by Yuma Wash, the opening of which comes into view between high walls of dirt. The drainage gets its name from the Yuma Mine, a galena (lead) mine far up the wash near Picture Rocks. The mine, which opened in 1885 but has been abandoned for decades, is famous for its minerals, particularly its striking brick red to yellow vanadinite and wulfenite crystals.

Across the floodplain to the west toward Silverbell Road is where in 1982 road construction crews discovered a thousand-year-old Hohokam village. Occupied from around AD 750 until as late as AD 1450, the settlement, it turned out, was huge. Archaeologists located more than a hundred dwellings, including pit houses and adobe-walled pueblos. Along with thousands of artifacts, excavators found rings, bracelets, and earrings carved from shells originating as far away as the Gulf of California and the Pacific Ocean.

A few years ago, I visited here with Courtney Rose, the project director from Old Pueblo Archaeology Center, a nonprofit educational and scientific organization, which was in the early phases of excavating. Volunteers wore masks and hats to screen buckets of dirt, raising a fine powder that clung to blue jeans and long-sleeved shirts. Courtney, her green shirt bulging in pregnancy and a Marshalltown trowel sticking out of her pocket, handed me a hand-sized piece of smooth, gray stone.

"What is it?" I asked her, feeling its single sharp edge.

"An agave knife," she answered. "For stripping the fiber from the agave leaves."

Jill, a young, energetic redhead in sunglasses and diamond nose-stud, offered to give me a tour. "Something really cool, if you want to take a look over here," she said, leading me to a Hohokam canal, which looked like paintings of dark brown and ocher semicircles in the wall of a backhoe trench. Next, weaving me around the workers and their bucket brigades, she showed me an excavated pit house, pointing out the round holes for the support posts. Jill explained that the Hohokam laid branches over a mesquite frame and covered everything with mud "from nearby puddling pits. We found the imprint of a basket in one, which is pretty exciting."

The archaeologists also found roasting pits with the charred bones of rabbits and deer, pieces of matted agave fiber, quartz crystals, and a rare turquoise pendant among all the shell and Tanque Verde red-on-brown pottery.

"This is the place to reconnect you with the past," Jill said.

The town of Marana is doing just that. Working with Old Pueblo Archaeology Center, the community plans to create a cultural heritage park along the Santa Cruz River. One component of the park will be the partial restoration of a nineteenth-century ranch established in 1878 by Juan and María Bojórquez and later homesteaded in 1900 by Feliberto Aguirre. Visitors will be able to tour the Bojórquez-Aguirre Ranch, which includes a home and a stonemasonry water tank, and explore its history with the aid of interpretive signs. Also in the plan is a signed Santa Cruz River Walk, another link of the Juan Bautista de Anza National Historic Trail. But the most encouraging part to me is the preservation of 90 percent of the Yuma Wash site. Here, researchers and craftspeople and volunteers will offer on-site tours of the digs. They will construct pit-house replicas, and they will display Hohokam artifacts to illustrate what it was like to live along the river when the first people farmed this valley.

That day, after touring the dig site, I walked between the high, narrow walls of Yuma Wash to where it spills back into the Santa Cruz, imagining intact Hohokam pots sticking out of the crumbling banks. A Bewick's wren trilled from the ragweed and saltbush, accompanied by the buzzing of a black-tailed gnatcatcher.

And there, charcoal gray against the white quartz, was a pottery sherd the size of my hand. It was thick and slightly curved, as if it were a fragment of an olla some ancient woman used to carry water from the river to her brush-and-mud home. I lifted it from the sand and felt its warmth against my palm. I raised it to my nose, thinking about the last person to touch that same piece of history. It smelled like life, redemption.

This evening I remember what the O'odham elder Daniel Preston said half a dozen years ago, standing in the rain next to that blue paloverde sapling we planted not too far downstream from here: how we've come full circle from where his ancestors once drew life from the Santa Cruz. The circle hasn't closed yet, I consider, seeing in the fading light other broken ollas at my feet. But there are many hands reaching, fingers extended, trying to fill the gap.

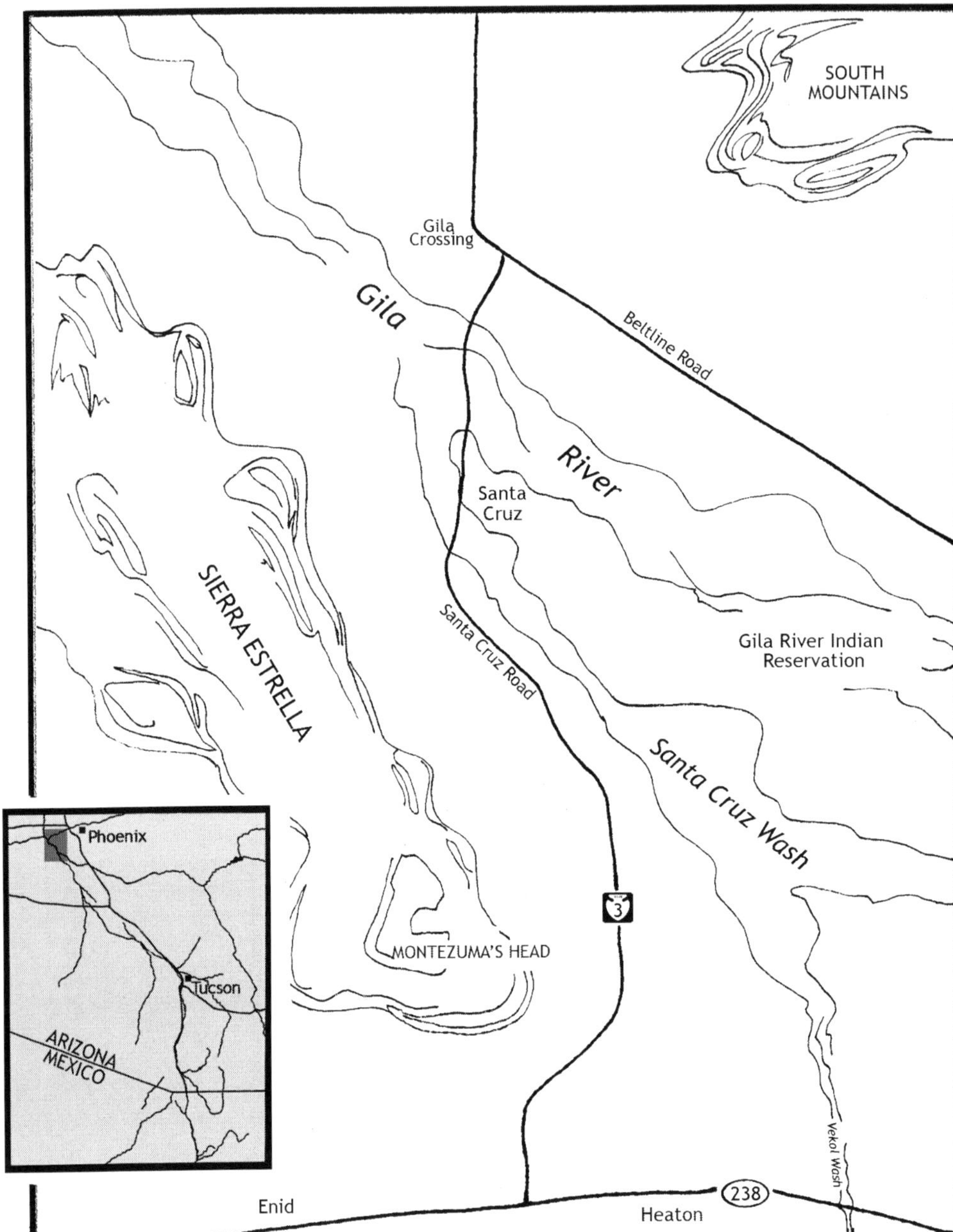

Confluences

9
Confluences

The Hobbit Hole Revisited: June, 150 River Miles

On a humid morning in late June as the monsoon season breaks upon us, Melissa and I meet with Kendall Kroesen at Tucson Audubon's Simpson restoration site thirty miles north of Tucson near the Santa Cruz River. Thunderheads billow in the south as moist winds from the Gulf of Mexico finally shove their way into southern Arizona. In minutes we're dripping with sweat, searching among the scrub for a familiar six-leaf-clover pattern in the ground. It's been nearly six years, and I'd forgotten how the rodent burrows suck your feet into the soft loamy soil, raising small clouds with every step.

Father Pedro Font must have been referring to this place when in late 1775 he traveled with the Anza expedition to California. The padre writes in his journal that the road from "El Tuquison" to the Gila River had little grass and even less water. He complains about bad footing due to the "squirrel mounds which abound" and how the trampling of people and horses lifted the choking dust.

"There are no trees in all this region," he writes on October 30. "Nor hardly a thing of value, for one sees only now and then a scrubby mesquite far in the distance. What is most abundant is the weed or shrub which they call the hediondilla [creosote bush], called in our country *la governadora*, and another scrubby and useless plant which, if the horses eat it, burns their mouths."

Two hundred and thirty-three years later, Kendall laments the slow recovery of the area, saying that it doesn't yet look like the "Sonoran Desert Garden of Eden" he'd been working toward and hoping for. But then he mentions Jeff Fehmi, a University of Arizona professor of rangeland restoration ecology who visited the site with his class recently. Jeff had pointed out that the place looks a good deal better, from the perspective of native vegetation and wildlife habitat, than it did when it was cotton fields.

"Jeff helped me to see from a longer-term perspective, and made me feel better about the progress we've made. Picture this as a retired fallow farm field, starting to grow weeds. It does show progress."

What I notice right away is the abundance of saltbush, large, recliner-sized shrubs trapping thickets of dead tumbleweed, which every lizard we stir up darts into. A line of cottonwood and willow marks the channel of the river to our south. Here and there stands a small acacia or paloverde, but the saltbush is amazing.

I have a photograph from 2002 that Kendall took of sixteen-year-old Melissa and me standing next to her hobbit hole, but I've neglected to bring it. It would have helped with all the saltbush. The photo shows the two of us posing behind her project, water squirting into her planting basins, surrounded by bare ground with one mesquite tree in the background. "We should look for the bubblers," Kendall says. "Some of them are still here.

As a way of locating the place, I'm picturing Melissa's creosote bush, which had tripled its size by the following year. But then Kendall says they had to narrow their list of species to those that thrived better. "Saltbush, screwbean mesquite, desert hackberry, wolfberry. The creosote and yucca didn't do as well, but partly because a flood in 2006 did some rearranging."

Despite the flood of two summers ago, we finally locate what we think is Melissa's hobbit hole. The teardrop shape of each basin is barely recognizable where each sweeps around the still-present PVC bubbler. Some of the basins are empty, including the one where we planted the yucca. The creosote, although it's as tall as my waist, looks dead. Its branches snap when I bend them. The saltbush, however, has taken over, crowding out three of the basins and more than half the circle. Kendall and I take pictures of Melissa, who looks like she's hiding behind the shrub.

We walk south through saltbush and burrobush toward the Santa Cruz River, flushing a coyote hidden in the willow thickets. I ask Kendall about the blue paloverde sapling blessed by Tohono O'odham elder Daniel Preston in 2002, and he leads us directly to it. The tree is no longer a bundle of dried sticks but an unfurling resurrection plant rising above us, bearing seedpods that rattle with life.

A Bell's vireo, that indicator bird of healthy riparian areas, sings from the cottonwoods. It's a husky melody that seemingly asks a question

Melissa Lamberton at hobbit hole six years later

and then answers it. "And there goes a summer tanager," Kendall says, raising his binoculars to his eyes as the bird flies across our path as red and luxuriant as a pomegranate. He thinks that white-tailed kites may be reproducing here. And another rare bird as well—the yellow-billed cuckoo. "Four summers ago we started hearing them, and now we think they're probably nesting." He pauses, and then scans the trees through his binoculars. "Nesting cuckoos would be a big coup for us because we didn't think this part of the river would be adequate for them."

It seems this part of the river is proving more than adequate, for birds and for more than birds. The river once may have been sold into bondage, its lifeblood drawn and poured out. But what I see here is that with just a little care and attention, some plantings of native vegetation and some free-flowing wastewater, the Santa Cruz can, indeed, redeem itself.

On our way back to the car, a huge desert iguana lumbers out of a burrobush and into the open to challenge us, as if Melissa and I were two enormous bipedal lizards with pale and weak scales who just trespassed into its territory. The animal flattens its hindquarters into the sand, and I'm immediately reminded of Pern, my daughters' pet green iguana who had a distinct personality bordering somewhere on rugged individualism and aloofness. This iguana doesn't care how big we are.

"He's doing push-ups," Melissa says when the blunt-headed, sand-dappled reptile begins bobbing the upper half of its fourteen-inch frame while at the same time mad-dogging us with one heavily lidded eye.

"Hey, cutie," Melissa says as she crouches closer, unruffled by the aggressive display. I watch the two of them in amazement, seeing the connection. *It's okay. We're only visiting,* she signals with her body, with her sharp blue eyes. *We know this is your home.*

The Simpson site lies at the far northwestern corner of a proposed national heritage area, what would be the next of only twenty-three such designations in the United States that define the American landscape. Shaped somewhat like the state of Illinois, the Santa Cruz NHA will encompass more than three thousand square miles of the upper and middle watershed of the Santa Cruz River. It will recognize that the Santa Cruz Valley is one of North America's oldest continuously cultivated communities, stretching into the past at least four thousand years, with the longest history of irrigated agriculture. From the canal-building Hohokam and their plots of squash to the modern well-drilling pecan farmers, the floodplain has nurtured humankind since the first immigrants crossed into the valley searching for a better life. It distinguishes a diversity of cultures, from the traditions of the Tohono O'odham saguaro-fruit harvesters and Spanish mission-builders to the Mexican and American frontiersmen and settlers. It describes a place rich with exposed geology, of earth stretched so thin its insides have burst out, of stones in midpause before being swept to the sea. It celebrates life as astonishing and inimitable as desert iguanas and borderland jaguars.

Santa Cruz, Arizona, 210 River Miles

A few days later, Karen and I drive sixty miles northwest of the Simpson site to the Gila Indian Reservation north of Maricopa, Arizona, where a sometimes channelized, sometimes meandering, multiple-branched Santa Cruz River joins the Gila River on its way west toward the Colorado River and then the Sea of Cortez. Hundreds of years ago, Father Pedro

Gila Indian church at the village of Santa Cruz, Arizona

Font called the confluence a "lagoon," which he described as formed by water "which runs into it from the plains during the rainy season, or from the Gila River itself when it overflows and wanders from its channel." Other historical writers described the confluence as a large marsh or slough. In fact, before William C. Greene's canal and subsequent flood-borne rerouting of the Santa Cruz in 1914–15, this lower reach of the river may have entered the Gila across an area as wide as twenty miles. The sheet flow of some floods apparently swept along both sides of the Sacaton Mountains north of Casa Grande.

These days, however, a flooding Santa Cruz River normally enters the Gila farther west over a smaller alluvial plain, one dotted with a tiny Indian hamlet appropriately named Santa Cruz.

Karen and I park a few miles southeast of the hamlet and directly south of the Gila River above a drainage marked "Santa Cruz Wash." A dust storm swept across the valley this evening, and now the sun glows Mars orange where it perches atop a backlit Montezuma Peak in the

Sierra Estrella. Blue-green saltbush rim a wide ditch pocked with rodent burrows, mesquite, and tamarisk.

"There's nothing on the map showing 'Santa Cruz Wash,'" Karen says, as we climb down a sloping bank through dried tumbleweeds. "The signs must be wrong. The map shows the river flowing farther south where the sign says 'Santa Rosa Wash.'" What we learn is that the map traces a Santa Cruz River that didn't exist a hundred years ago.

Although most of it is now buried under farmland, the Santa Cruz Wash at this place is a remnant of one of the main braided paths the former river took before Greene's canal cut off and channelized it, effectively restricting runoff to the Santa Cruz Flats. Before the diversion, floodwaters apparently flowed in a more northerly direction, inundating areas that are now covered by the communities of Eloy, Toltec, and Casa Grande.

It's the Santa Rosa Wash that now carries the river to the Gila. The new path begins where the gun-slinging, ditch-digging Colonel Greene carved his thirteen-mile canal southwest of Picacho Peak, intending to divert water from the Santa Cruz River westward into a reservoir, which floods destroyed along with his levees. From the abandoned reservoir, Greene's Canal becomes Greene Wash, turns northwest, crosses I-8, and joins the Santa Rosa Wash about a mile northeast of the town of Stanfield. Here, the Santa Rosa Wash enters some croplands and spreads out into multiple channels north of Maricopa. These branches then collect the Santa Cruz Wash flowing east-west between the Casa Grande and Sacaton mountains, and the broad, sinuous, and now reconnected river sweeps through the village of Santa Cruz to finally meet the Gila River.

From its headwaters in the San Rafael Valley, its loop into Mexico and return to Arizona, the Santa Cruz River has traveled more than two hundred miles. And I have seen it all.

In *The Lessening Stream*, Michael Logan says that the tiny community of Santa Cruz was wiped out by the 1993 flood. I imagine the 1993 flood is just the most recent of many to devastate the village. Even Father Kino, who visited here in 1697, mentions meals without pinole due to washed-out fields. Witnessing the place where the Santa Cruz dies, I also expect to see the village's death—all of it in exquisite detail, like individual feathers dusted on glass after a bird's impact.

The village rests quietly in the Y formed where the two rivers meet, inside a mile-and-a-half-wide tongue of sediments crowded with

Santa Cruz and Gila rivers near confluence

mesquite and tamarisk trees at the base of the outwash bajadas of the Sierra Estrella on the west and the perpendicular finger of South Mountain on the east. This is the place of the historic Gila Crossing, a funnel of rivers and humanity, the latter at the mercy of the whims of the former.

Too little water, too much water. There has always been death here. But it's a death that only gives dimension to life as shadows give form and depth to what's featureless.

Rivers are such a powerful source for symbolism.

I like to think of the Santa Cruz River as a metaphor representing, among other things, my own life. Between its inconsequential birth and inevitable death, the river ventures across an unforeseeable landscape, reverses directions once, while slowly losing ground along its way. Sometimes it flows wildly, tossing rocks and earth and uprooting trees and homes while testing new aftermaths. Sometimes it sinks out of sight. Mostly it smells like sewage, the dung of life, of rebirth.

People are like water on the land. Some flow through it, adapting and adjusting to its hard, unmovable terrain. Others fight the land, trying to carve their own path. And, although the land may be indifferent to the path we take across it, the land affects the path all the same, as it affects the path water takes across it.

I write these final thoughts as the first thunderstorm of the monsoon season sends a second wave of pounding rain across this small patch of desert. Earlier, after I crossed a raging Santa Cruz River on my way home, I found the spadefoots were calling, their urgent bleats rising out of the San Juan Wash behind the house. Now both arroyos on either side of this place gush with muddy runoff, effectively trapping me here. Not that I mind.

This monsoon evening I'm alone. Karen lives ninety miles southeast of me in Lower Stump Canyon of the Huachuca Mountains, a few miles north of our border with Mexico. In the next few weeks I will join her there, as we begin to relocate to southern Cochise County and the quaint mining town of Bisbee where Karen now works. Jessica and Melissa have their jobs and degree programs in Tucson, the two living in a small and charming 1940s Spanish casita near midtown. Kasondra teaches grade school in Phoenix. Our family collective is finally breaking up as our daughters discover tributary paths of their own. I'm amazed it lasted this long—well into the girls' college years—but I take comfort in knowing that my family, like this river we've lived alongside most of our lives, eventually must become something greater than itself.

Our six-hundred-square-foot home on five acres of Sonoran upland above San Juan Wash seems too large for me now, even though we dismantled the girls' room and their triple-bunked beds some years ago when they left for college and filled the space with desks and bookshelves and file cabinets. The kitchen table we crowded around is an empty spot on the bare tile; I eat meals staring out the window above the sink, looking vacantly through a foothill paloverde where my hummingbird feeder has always hung but no longer does.

It will be hard for me to leave this place, especially when I think about what Karen and my girls have experienced here over the decades, the snow-capped saguaro cacti, the tadpole rescues and toad-rearing, the many pet graves. In only the second time since Karen was a child, a tenacious night-blooming cereus, the only one on the property, will bloom this summer. Its single bud fattens even now with the rain.

It will be really difficult when I recall all those evenings following a monsoon thunderstorm when something truly remarkable entered our lives, like on this very evening when a hopelessly confused, bubble-eyed Couch's spadefoot slipped through the cat door and hopped across my fake saltillo tile.

But where one river may run its course, others rise out of bedrock. There's another river that runs through my life now, and I find encouragement and pleasure in the possibility of getting to know it. Like the Santa Cruz, the San Pedro River also flows north into southern Arizona from Mexico, only without first making a significant course correction (a comforting thought). Its banks are thick with willows and cottonwoods, its waters tannin-rich and aromatic. It is a river not unlike Tucson's Santa Cruz a hundred and fifty years ago.

If I have my way and fortune comes to me, I will live on one of its tributaries, perhaps in the oak-draped canyon where Banning Creek creases the western haunches of the Mule Mountains below the Bisbee "Tunnel of Love" and the famously misplaced "Continental Divide." There, I might spend my days hiking the old Mule Pass Road, thinking about the dry rivers in my life, and how they have shaped me and continue to do so as they shape this grand and astonishing landscape.

Acknowledgments

For holding my hand through the eight years in writing this book, my deepest and sincere gratitude goes to my writing compadres: Elizabeth Bernays, Tony Leubbermann, Mac Hudson, Richard Shelton, Erec Toso, Alison Hawthorne Deming, Fenton Johnson, Alan Weisman, Madeline Kiser, Ralph Hager, Steve Gladish, John Mead, Jerry Marzinsky, Gillian Haines, Dawn Sellers, Jennifer Schneider, Barbara Stahura, Walker Thomas, Valarie James, Rita Henry, Marilyn Gustin, Jimmye Hillman, Brenda Hillman, Robert Hass. Thank you for your criticism and encouragement and the many words read and re-reread.

To Julie St. John, for all those coffee-inspired discussions at Raging Sage.

To Zay Hartigan, for your most excellent rescue.

To the brilliant people of Tumacácori National Historic Park, Tucson Audubon Society, Sky Island Alliance, Arizona State Museum, Arizona Historical Society, Tubac Historical Society, for telling me your stories. I can only hope I do them justice in these pages.

To the tribal council of the Tohono O'odham Nation, San Xavier District, for, by a vote of 5–1, allowing me to hike across your lands.

To the many people—scientists, historians, researchers, teachers, ranchers and wilderness enthusiasts, bird-watchers and river aficionados—for your patience and understanding with all my ridiculous questions, whether we met on the river, in some dusty museum, or only through e-mail:

Mary Bingham, Don Garate, Grant Hilden, Sonja Macys, Acacia Berry, Gita Bodner, Mike Quigley, Sergio Avila, Trevor Hare, Bernardo Espinoza, Blanca Ophelia Galaz, Carlos Robles, Marge and Ron Izzo, Bill McClain, Daniel Preston and Renee Red Dog, Kendall Kroesen, Rodd Lancaster, Ann Phillips, Michael Bodin, Jimmy Straley, Mark Briggs, Scott Wilbor, Michael Bissontz, Charles Riggs, Michael Waters,

Tim Snow, Randy Babb, Dan Koskuba, Homer Thiel, Jonathan Mabry, Danita Rios, Phil Rosen, Hector and Mickie Soza, Dan Daly, Gerald Talen, Rick Collins, Courtney Rose, Diana Hadley, Peter Warshall, Janine and Tim McCabe.

To Jessica Lamberton, for all your expert help with the maps.

And to my editor, Patti Hartmann, and all those at the University of Arizona Press—Kathryn Conrad, Holly Schaffer, Leigh McDonald, Kristen Buckles, Barbara Yarrow, and many others—for making this book possible.

Various excerpts of this book have appeared in Sky Island Alliance's newsletter, *Restoring Connections*, the *Los Angeles Times*, *The Fourth River*, *Walking Rain Review*, *Qarrtsiluni*, and the anthologies *Art in Wilderness: Tumacácori Highlands*, *Stories from the Other Side*, fifth edition, and *Living Now: Mending Our Place in the Natural World*.

This project was supported by a generous grant from the Arizona Commission on the Arts.

Timeline

1536 Estevanico (Black Steven) and Cabeza de Vaca in Sonora. Chapter 2

1539 Fray Marcos de Niza and Estevanico enter San Rafael Valley. Chapter 2

1540 Francisco Vásquez de Coronado enters Arizona. Chapter 2

1637 Juan Munguia Villela and General Pedro de Perea come to Santa Cruz, Sonora. Chapter 3

1645 Father Eusebio Francisco Kino born in Segno, Italy. Chapter 3

1681 Father Kino arrives at Vera Cruz, Sonora. Chapter 3

1687 Father Kino establishes first mission, Nuestra Señora de los Dolores, on March 13. Chapter 3

1691 Father Kino establishes visita at San Lázaro and visits Tumacácori and Guevavi. Chapters 3, 4

1692 Father Kino establishes mission at San Xavier del Bac. Chapter 7

1693 Father Kino establishes mission at Santa Cruz, Sonora (Santa María Suamca). Chapter 3

1697 Father Kino visits Pima tribes at Gila River. Chapter 9

1698 Father Kino establishes a visita at Tucson. Chapter 8

1701 Father Kino establishes mission at Guevavi. Chapter 4

1701 Father Francisco Gonzalvo at San Xavier. Chapter 8

1711 Father Kino dies at Magdalena. Chapter 4

1726 Father Joseph Agustín visits Tubac (Tchoowaka). Chapter 5

1732 Father Ignacio Xavier Keller comes to Santa Cruz, Sonora (Santa María Suamca). Chapter 3

1732 Father Johann Baptist Grazhoffer reestablishes Guevavi mission. Chapter 4

Tubac becomes visita to Guevavi mission. Chapter 5

1736 Juan Bautista de Anza born in Sonora. Chapter 5

1750s Father Joseph Garrucho builds mission at Guevavi. Chapter 4

1751 Pima Indian Revolt. Chapters 4, 5

1752 Presidio at Tubac established under Captain Juan Tomás Belderrain. Chapter 4

1756 Calabasas becomes a visita under Father Francisco Pauer. Chapter 4

1767 Jesuit expulsion. King Charles III banishes Jesuits from Spanish lands. Chapters 3, 4

Franciscans arrive in Sonora. Chapter 4

1768 Apaches burn the church at Santa Cruz, Sonora (Santa María Suamca). Chapter 3

1771 Father Francisco Garcés begins building project at Tucson. Chapter 8

1775 August 20, Tucson is born when Colonel Hugo O'Conor orders the Tubac presidio to Tucson. Chapter 4

Juan Bautista de Anza leads colonizing expedition through Arizona. Chapter 4

Father Pedro Font writes about lower Santa Cruz River. Chapter 9

1776 The presidio of Santa Cruz de Terrenate established at the San Pedro River. Chapter 3

1776 Tubac presidio transfers to Tucson. Chapter 4

1777 Father Juan Bautista Belderrain at Mission San Xavier. Chapter 7

1778 De Anza becomes governor of New Mexico. Chapter 5

1778 Father Francisco Garcés begins Convento at Tucson. Chapter 8

1779 Spanish abandon the presidio Santa Cruz de Terrenate. Chapter 5

1781 Tubac abandoned. Chapter 5

1781 Father Francisco Garcés martyred at Yuma. Chapter 8

1782 "May Day" Apache attack at Tucson presidio. Chapters 7, 8

1783 Father Belderrain begins construction of present mission at San Xavier. Chapter 7

1787 The presidio of Santa Cruz de Terrenate moves to Santa Cruz, Sonora (Santa María Suamca). Chapter 3

1787 Tubac receives detachment of Pima Indians, San Rafael Company. Chapter 5

1788 De Anza dies at Arizpe. Chapter 5

1789 Toribio Otero receives Spanish land grant at Tubac. Chapter 6

1790s Father Juan Bautista Llorens begins Convento expansion. Chapter 8

1797 Father Juan Baptista Llorens completes San Xavier Mission. Chapter 7

1800 Franciscans under Fray Narciso Gutierrez begin construction of present church at Tumacácori. Chapter 5

1814 Father Juan Bautista Llorens departs Bac and dies in 1815. Chapter 8

1821 September 27: Mexico wins independence from Spain. Chapter 4

December 15: Ignacio and Tomás Ortiz receive La Canoa Spanish land grant. Chapter 6

1822 Present church at Tumacácori completed. Chapter 5

1825 Ramón Romero receives San Rafael Spanish land grant. Chapter 2

1828 Mexican government orders all Spanish out of country. Chapter 5

1831 Father Rafael Diaz departs Tucson for Ímuris. Chapter 8

1834 Mexican government decrees end to mission system. Chapter 4

1837 Manuel María Gándara becomes governor of Sonora first time. Chapter 4

1847 Felix Telles, "Mickey Free," is born in Santa Cruz, Sonora. Chapter 3

1848 Treaty of Guadalupe Hidalgo. Chapter 2

Tumacácori mission abandoned. Chapter 5

1850 Fifteen-year-old Inez Gonzales is captured by Apaches. Chapter 3

1850s Alfred and William Rowlett dam Santa Cruz to create Silver Lake. Chapter 8

1851 John Russell Bartlett enters San Rafael Valley. Chapters 2, 3

1852 Bartlett visits Tubac and Tucson. Chapters 6, 8

1854 Gadsden Purchase is ratified. United States agrees to pay Mexico $10 million for 45,535 acres of land below the Gila River from the Rio Grande to the Colorado River. Chapter 2

Charles Poston at Tubac. Chapter 5

Gándara's Calabasas church becomes ranch house. Chapter 4

1856 Ignacio Pesqueira becomes governor of Sonora. Chapter 4

Elias C. Brevoort becomes Tucson's first postmaster. Chapter 6

Solomon Warner arrives at Tucson. Chapter 8

1858 Samuel Hughes arrives at Tucson. Chapter 8

1859 Elias Pennington at Calabasas. Chapter 4

Arizona's first newspaper begins publication at Tubac. Chapter 5

1860 Larcena Pennington survives Apache attack. Chapter 4

William S. Grant builds flour mill at Silver Lake. Chapter 8

1860s Fort Buchanan abandoned by U.S. Army. Chapter 4

1861 Mickey Free is captured by Apaches. Chapter 3

1862 Jefferson Davis creates Confederate Territory of Arizona. Chapter 6

Battle of Picacho Pass. Chapter 6

1864 James Lee arrives at Tucson. Chapter 8

1867 Sabino Otero at Tubac. Chapter 6

1870 Seven sisters of St. Joseph Carondelet arrive at Tucson. Chapter 8

1874 President Ulysses S. Grant establishes Papago (Tohono O'odham) Reservation at San Xavier Mission. Chapter 7

Solomon Warner builds flour mill under Sentinel Peak. Chapter 8

1876 Tomás Ortiz sells La Canoa Ranch to Frederick Maish and Thomas Driscoll. Chapter 6

1878 Gándara sells Calabasas to Charley Sykes. Chapter 4

Gándara dies. Chapter 4

1879 James Kilroy Brown at ranch in Sahuarita. Chapter 6

1880 Railroad arrives at Tucson. Chapter 8
St Mary's Hospital dedicated. Chapter 8

1880s Lee's Pond becomes Silver Lake. Chapter 8

1881 Manuel H. Amado evicted from home. Chapter 6

1882 Colin Cameron arrives in Tucson. Chapter 2
Nogales founded. Chapter 4

1883 Colin Cameron forms San Rafael Cattle Company. Chapter 2

1884 Luttrell becomes Lochiel, Arizona. Chapter 2

1886 Geronimo's band attacks Al Peck and family. Chapter 5

1888 Samuel Hughes digs headcut in Santa Cruz River. Chapter 8

1890 Santa Cruz River becomes "Tucson Arroyo." Chapter 8

1891 Solomon Warner closes mill. Chapter 8

1894 U.S. Court of Private Land Claims voids Spanish and Mexican land grants. Chapter 2

1897 Colonel William C. Greene kills Jim Burnett at Tombstone. Chapter 2

1899 Colonel Greene forms Cobre Grande Copper Company in Sonora. Chapter 2

1903 Colonel Greene at San Rafael Ranch. Chapter 2

1908 President Roosevelt establishes Tumacácori National Monument. Chapter 5

1911 March 6: Colin Cameron dies in Tucson. Chapter 2
July 31: Colonel Greene dies. Chapter 2

1912 Levi H. Manning purchases the 17,203-acre San Ignacio de la Canoa Spanish land grant. Chapter 6
Menlo Park founded. Chapter 8

1914 Sabino Otero dies. Chapter 6
Teofilo Otero inherits Otero ranch empire. Chapter 6

1916 Intercontinental Rubber Company purchases northern half of La Canoa from Levi Manning. Chapter 6

1920 Amadoville becomes Amado. Chapter 6

1921 Howell Manning takes over La Canoa Ranch operations. Chapter 6

1931 Santa Cruz, Sonora, become a municipality. Chapter 3

1938 Rex Hamaker founds Rex Ranch Resort and dude ranch. Chapter 6

1939 Al Peck dies in Nogales. Chapter 5

1941 Teofilo Otero dies. Chapter 6

1950 Farmers Investment Company purchases northern portion of La Canoa Ranch. Chapter 6

1950s Convento in Tucson bulldozed. Chapter 8

1951 Howell Manning Jr. is killed by a truck driver on December 21. Chapter 6

Felix Lucero dies. Chapter 8

1956 Tucson architect Josias Joesler dies. Chapter 6

1957 Agua Linda sold to Arthur Loew Jr. Chapter 6

1958 Tubac presidio becomes Arizona's first state park. Chapter 5

1959 Tubac Golf Resort begins construction. Chapter 6

1966 Father Eusebio Kino's bones discovered at Magdalena, Sonora. Chapter 4

1968 Duval Corporation purchases portion of La Canoa Ranch. Chapter 6

1970s I-19 begins construction. Chapter 6

1981 Desert Survivors Native Plant Nursery founded. Chapter 8

1982 Garden of Gethsemane, Felix Lucero Park, created. Chapter 8

1990 Congress authorizes Juan Bautista de Anza National Historic Trail. Chapter 5

Tumacácori National Monument becomes Tumacácori National Historical Park. Chapter 5

1998 Nature Conservancy acquires San Rafael Ranch. Chapter 2

Sweetwater Wetlands opens to public. Chapter 8

2002 December 22, blessing ceremony for the lower Santa Cruz restoration project. Chapter 1

2003 Kino Heritage Fruit Trees Project started. Chapter 8

2007 May 19, re-created Presidio de San Agustín del Tucson opens to public. Chapter 8

Selected Bibliography

Bartlett, John R. *Personal Narrative of Explorations and Incidents in Texas, New Mexico, California, Sonora, and Chihuahua, 1850–1853*. 1884. Reprint, Chicago: Rio Grande Press, 1965.

Betancourt, Julio Luis. "Tucson's Santa Cruz River and the Arroyo Legacy." PhD diss., University of Arizona, Tucson, 1990.

Bleser, Nicolas J. *Tumacácori: From Rancheria to National Monument*. Tucson: Southwest Parks and Monuments Association.

Bolton, Herbert Eugene. *Anza's California Expeditions*. Berkeley: University of California Press, 1930.

———. *The Padre on Horseback: A Sketch of Eusebio Francisco Kino, S.J., Apostle to the Pimas*. 1932. Reprint, Chicago: Loyola Press, 1963.

———. *Rim of Christendom: A Biography of Eusebio Francisco Kino, Pacific Coast Pioneer.* New York: Macmillan, 1936.

Brandt, Herbert. *Arizona and Its Bird Life: A Naturalist's Adventures with the Nesting Birds on the Deserts, Grasslands, Foothills, and Mountains of Southeastern Arizona*. Cleveland: Bird Research Foundation, 1951.

Briggs, Mark K. *Riparian Ecosystem Recovery in Arid Lands: Strategies and References*. Tucson: University of Arizona Press, 1996.

Brown, James Cabell. *Calabasas: Or, Amusing Recollections of an Arizona City*. 1892. Reprint, Tucson: Arizona Silhouettes, 1961.

Brownell, Elizabeth R. *Peck Canyon through the Centuries*. Unpublished manuscript. Tubac Historical Society, Tubac, Arizona, 1977.

———. *They Lived in Tubac*. Tucson: Westernlore Press, 1986.

Bryne, Leo G., and Sister Alberta Cammack, C.S.J. "Heritage: The Story of St. Mary's Hospital, 1880–1980." 1980. http://parentseyes.arizona.edu/carondelet/heritage_toc.html.

Casanega, Thomas David. "The Life History of T. D. Casanega." Unpublished papers, 1883–1935. Arizona Historical Society, Tucson.

Cleland, Robert Glass. *This Reckless Breed of Men: The Trappers and Fur Traders of the Southwest*. New York: Knopf, 1950.

Clizbe, George A. *This Is Green Valley.* Tucson: Shandling Lithographing Co., 1971.

Cremony, John C. *Life among the Apaches.* 1868. Reprint, Lincoln: University of Nebraska Press, 1983.

Czaplicki, Jon D., and James D. Mayberry. "An Archaeological Assessment of the Middle Santa Cruz River Basin." University of Arizona Archaeological Series No. 164. Tucson: University of Arizona Press, 1983.

Deaver, William L., and Carla R. Van West, eds. *El Macayo: A Prehistoric Settlement in the Upper Santa Cruz River Valley.* Tucson: Statistical Research, Inc, 2001.

Dobyns, Henry F. *Spanish Colonial Tucson: A Demographic History.* Tucson: University of Arizona Press, 1976.

———. *Tubac through Four Centuries: An Historical Resume and Analysis.* Arizona State Parks Board, 1959. Reformatted for electronic Web version by Tubac Presidio State Historical Park, August 1995. http://parentseyes.arizona.edu/tubac/index.html.

Doelle, William H. "Human Use of the Santa Cruz River in Prehistory." Paper presented at the Biennial Conference of the American Society for Environmental History, Tucson, Arizona, April 14–18, 1999.

Eckhart, George B., and James S. Griffith. "Temples in the Wilderness." Historical Monograph, No. 3. Tucson: Arizona Historical Society, 1975.

Engelhardt, Zephyrin. *The Franciscans in Arizona.* Harbor Springs, MI: Holy Childhood Indian School, 1899.

Fontana, Bernard L. "Biography of a Desert Church: The Story of Mission San Xavier del Bac." Tucson: The Smoke Signal, Tucson Corral of the Westerners, 1961, 1996.

———. "Calabazas of the Rio Rico." Tucson: The Smoke Signal, Tucson Corral of the Westerners, 1971.

Giffords, Gloria Frasier. *Sanctuaries of Earth, Stone, and Light: The Churches of Northern New Spain, 1530–1821.* Tucson: University of Arizona Press, 2007.

Glennon, Robert. *Water Follies: Groundwater Pumping and the Fate of America's Fresh Waters.* Washington, D.C.: Island Press, 2002.

Granger, Bird Howell. *Arizona's Names (X Marks the Place): Historical Names and Places in Arizona.* Tucson: Treasure Chest Publications, 1983.

Greenwald, Dawn M., and Dennis Gilpin. "Archaeological Overview of the Santa Cruz River Valley." Unpublished report, Arizona State Land Department, 1996.

Halpenny, Leonard C. "Ground-Water Resources within the San Xavier Indian Reservation and Proposals Relating to Leases for Development of Ground Water." Tucson: Water Development Corporation, 1962.

———. "Review of the Hydrology of the Santa Cruz Basin in the Vicinity of the Santa Cruz-Pima County Line." Paper given at the First Annual Conference of the Arizona Hydrological Society, Phoenix, Arizona, September 16, 1988.

Hardy, Robert William H. *Travels in the Interior of Mexico, in 1825, 1826, 1827 and 1828.* 1829. Glorieta, NM: Rio Grande Press, 1977.

Hastings, James R. *Vegetation Change and Arroyo Cutting in Southeastern Arizona during the Past Century: An Historical Review.* 2002. http://alic.arid.arizona.edu/sonoran/documents/hastings/alc_arroyo.html

Heald, Weldon F. *Sky Island—A Retreat into One of Nature's Last Strongholds, the Spectacular Chiricahua Mountains.* D. Van Nostrand Company, 1967. Reprinted as *The Chiricahua Mountains.* Tucson: University of Arizona Press, 1975.

Kessel, John L. *Friars, Soldiers, and Reformers: Hispanic Arizona and the Sonora Mission Frontier, 1767–1856.* Tucson: University of Arizona Press, 1976.

———. *Mission of Sorrows: Jesuit Guevavi and the Pimas, 1691–1767.* Tucson: University of Arizona Press, 1970.

Logan, Michael F. "Head-cuts and Check-dams: Changing Patterns of Environmental Manipulation by the Hohokam and Spanish in the Santa Cruz River Valley, 200–1820." *Environmental History* 4, no. 3 (1999): 405–30.

———. *The Lessening Stream: An Environmental History of the Santa Cruz River.* Tucson: University of Arizona Press, 2002.

Mabry, Jonathan B. "Three Thousand Years of Irrigation in a Riverine Oasis." *Archaeology Southwest* 15, no. 2 (2001): 14–15.

Matson, Daniel S., and Bernard L. Fontana, eds. *Friar Bringas Reports to the King: Methods of Indoctrination on the Frontier of New Spain, 1796–97.* Tucson: University of Arizona Press, 1977.

McCarty, Kieran. *Desert Documentary.* Tucson: Arizona Historical Society, 1976.

McPhee, John. *Basin and Range.* New York: Noonday Press, 1980.

Morner, Magnus. "The Expulsion of the Jesuits from Spain and Spanish America in 1767 in Light of Eighteenth-Century Regalism." *The Americas* 23, no. 2 (1966): 156–64.

Nentvig, Juan. *Rudo Ensayo: A Description of Sonora and Arizona in 1764.* Translated and edited by Alberto F. Pradeau and Robert R. Rasmussen. Tucson: University of Arizona Press, 1980.

Officer, James. *Hispanic Arizona, 1536–1856.* Tucson: University of Arizona Press, 1987.

Parker, John T. C. "Channel Change on the Santa Cruz River, Pima County, Arizona, 1936–1986." Tucson: U.S. Geological Survey Water-Supply Paper 2429, 1995.

Peck, Marguerite O'Brien. *In the Memory of a Man.* Unpublished manuscript, Special Collections, University of Arizona Library, Tucson; Tubac Historical Society, Tubac, 1966.

Pfefferkorn, Ignacio. *Sonora: A Description of Province.* Albuquerque: University of New Mexico Press, 1949.

Pickens, Buford, ed. *The Missions of Northern Sonora: A 1935 Field Documentation.* Tucson: University of Arizona Press, 1993.

Pima County Parks and Recreation. "The Canoa Ranch Master Plan, Background Report." Tucson: Pima County Parks and Recreation, 2006.

———. "The Canoa Ranch Master Plan, Final Report." Tucson: Pima County Parks and Recreation, 2007.

Polzer, Charles W. *Kino Guide II: A Life of Eusebio Francisco Kino, S.J. Arizona's First Pioneer and a Guide to His Missions and Monuments.* Tucson: Southwestern Mission Research Center, 1982.

Pumpelly, Raphael. "Pumpelly's Arizona"; an excerpt from *Across America and Asia*, by Raphael Pumpelly, comprising those chapters which concern the Southwest. Tucson: Palo Verde Press, 1965.

Radding, Cynthia. *Wandering Peoples: Colonialism, Ethnic Spaces, and Ecological Frontiers in Northern Mexico, 1700–1850.* Durham, NC: Duke University Press, 1997.

Riggs, Charles R., and Rein Vanderpot. "Archaeological Testing Report and Treatment Plan for the Lone Mountain Ranch Land Exchange in the San Rafael Valley, Arizona." Technical Report 97-4. Tucson: Statistical Research, Inc., 1997.

Scott, Sandra, ed. *The Pimería Alta.* Tucson: Southwestern Mission Research Center, Arizona State Museum, University of Arizona, 1996.

Sheridan, Thomas E. *Los Tucsonenses: The Mexican Community in Tucson, 1854–1941.* Tucson: University of Arizona Press, 1986.

Sonnichsen, C.L. *Colonel Greene and the Copper Skyrocket.* Tucson: University of Arizona Press, 1974.

———. *Tucson: The Life and Times of an American City.* Norman: University of Oklahoma Press, 1982.

Spicer, Edward H. *Cycles of Conquest: The Impact of Spain, Mexico, and the United States on the Indians of the Southwest, 1533–1960.* Tucson: University of Arizona Press, 1962.

———. "The Military History of the Yaquis from 1867 to 1910: Three Points of View." 2001. http://usaic.hua.army.mil/History/Html/spicer.html.

Spring, John A. *John Spring's Arizona.* Edited by A. M. Gustafson. Tucson: University of Arizona Press, 1966.

Tellman, Barbara, and Richard Yarde. *A Historical Study of the Santa Cruz River: Background Information for Determination of Navigability of the River at the Time of Arizona Statehood, 1912.* Unpublished report, Arizona State Land Department, 1996.

Thiel, J. Homer. "Down by the River: Archaeological and Historical Studies of the León Family Farmstead." Archaeological Papers, No. 32. Center for Desert Archaeology, August 2005.

Trimble, Marshall. *Roadside History of Arizona.* Missoula, MT: Mountain Press Publishing Company, 1986.

Van West, Carla R., and Stephanie M. Whittlesey. *Cultural Resource Management Plan of the Fairfield Canoa Ranch Property.* Vol. 1: *Background and Research Design for Prehistoric Archaeological Resources.* Tucson: Statistical Research Institute, 1997.

Waters, Michael R. "Disaster or Catastrophe: Human Adaptation to High- and Low-Frequency Landscape Processes—A Reply to Ensor, Ensor, and Devries." *American Antiquity* 68 (2003): 400–405.

———. "Holocene Alluvial Geology and Geoarchaeology of the San Xavier Reach of the Santa Cruz River, Arizona." *Geological Society of America Bulletin* 100, no. 4 (1988): 479–91.

———. "The Impact of Fluvial Processes and Landscape Evolution on Archaeological Sites and Settlement Patterns along the Santa Cruz River, Arizona." *Geoarchaeology* 3, no. 3 (1988): 205–19.

Wayland, Jane Abigail. "Experiment on the Santa Cruz: Colin Cameron's San Rafael Cattle Company, 1882–1893." Master's thesis, University of Arizona, 1964.

Webb, Robert H., Stanley A. Leake, and Raymond M. Turner. *The Ribbon of Green: Change in Riparian Vegetation in the Southwestern United States.* Tucson: University of Arizona Press, 2007.

Weisman, Alan. *La Frontera: The United States Border with Mexico.* San Diego: Harcourt, Brace, Jovanovich, 1986.

Whittlesey, Stephanie M. "Culture History: Prehistoric Narratives for Southern Arizona." *Cultural Resource Management Plan of the Fairfield Canoa Ranch Property.* Vol. 1: *Background and Research Design for Prehistoric Archaeological Resources.* Tucson: Statistical Research Institute, 1997.

Wilbor, Scott. *Avian Habitat Conservation Plan, U.S. Upper Santa Cruz River Riparian Corridor, Santa Cruz County, Arizona.* Tucson: Tucson Audubon Society, 2005.

Willey, Richard R. "La Canoa: A Spanish Land Grant Lost and Found." Tucson: The Smoke Signal, Tucson Corral of the Westerners, 1979.

Wood, Michelle Lee, P. Kyle House, and Philip A. Pearthree. "Historical Geomorphology and Hydrology of the Santa Cruz River." Unpublished report, Arizona State Land Department, 1996.

Wormser, Richard. *Tubac.* Tubac: Tubac Historical Society, 1975.

Zoontjens, Linda. "Brief History of the Yaqui and Their Land." 2001. http://sustainedaction.org/Explorations/history_of_the_yaqui.htm.

Index

About the Author

When Ken Lamberton published his first book *Wilderness and Razor Wire*, the *San Francisco Chronicle* called it "entirely original: an edgy, ferocious, subtly complex collection of essays." The book won the 2002 John Burroughs Medal for outstanding nature writing. He has published four books and more than a hundred articles and essays in places like the *Los Angeles Times*, *Arizona Highways*, the *Gettysburg Review*, and the *Best American Science and Nature Writing 2000*. In 2007, he won a Soros Justice Fellowship for his fourth book, *Time of Grace: Thoughts on Nature, Family, and the Politics of Crime and Punishment* (University of Arizona Press, 2007). He holds degrees in biology and creative writing from the University of Arizona and lives with his wife in an 1890s stone cottage near Bisbee. Visit the author's website at www.kenlamberton.com.